TRANSFORM METHOD IN
LINEAR SYSTEM ANALYSIS

McGRAW-HILL ELECTRICAL AND ELECTRONIC ENGINEERING SERIES

FREDERICK EMMONS TERMAN, *Consulting Editor*
W. W. HARMAN AND J. G. TRUXAL,
Associate Consulting Editors

TRANSFORM METHOD IN LINEAR SYSTEM ANALYSIS

JOHN A. ASELTINE

Member of the Technical Staff
Space Technology Laboratories
Lecturer in Engineering
University of California at Los Angeles

McGRAW-HILL BOOK COMPANY, INC.

New York   Toronto   London

1958

TRANSFORM METHOD IN LINEAR SYSTEM ANALYSIS

Copyright © 1958 by the McGraw-Hill Book Company, Inc. Printed in the United States of America. All rights reserved. This book, or parts thereof, may not be reproduced in any form without permission of the publishers. *Library of Congress Catalog Card Number* 58-8038

8 9 10 11 12 13 14 15 – MP – 1 0 9 8 7

02389

To Jane
and
Mark, Ann, Jean, and Martha

PREFACE

This book is about the application of integral transforms to the analysis of physical systems which can be described by linear differential equations. It is an engineering book, and most of the mathematics it contains is presented from the engineer's rather relaxed point of view regarding rigor.

The purpose of this book is, first, to show what the transform method is and how it is applied to a variety of engineering problems. Second, and perhaps more important, I have tried to show how the transform method can give an *understanding* of physical phenomena. To this end considerable space is devoted to the concepts of system function, frequency response, impulse response, and the like. Finally, I have tried to bring together material related to the analysis of linear systems and to show as clearly as possible the relationships between the various methods.

This book uses a mathematical technique—the transform method—to solve as well as explain engineering problems. The approach is heuristic —that is, arguments about the mathematics are intuitive rather than logically compelling. The justification for this approach lies in the way an engineer uses mathematics. He wants to use it to get answers as quickly as possible, and he wants it to aid him in understanding the physical world. He is willing to make occasional mistakes due to lack of rigor if these first objectives are achieved. He knows that the mistakes will be discovered in the measurements he must make to check his theoretical results.

In this book I have tried to make a compromise between rigor and intuition consistent with the engineering point of view. In places where the intuitive mathematics leads to difficulties, an explanation of the source of trouble is given, together with reference to rigorous treatment elsewhere. In general I have taken the attitude that the use of non-rigorous mathematics is justified if it leads to better understanding and more direct solution of engineering problems, and if the user has knowledge of its limitations.

It is assumed that the reader will have an acquaintance with elementary differential equations. Prior exposure to electricity and mechanics is also required, although the equations describing physical systems used as examples are derived in the text.

The first half of the book deals with the Laplace transform[1] and its application to problems in electricity and mechanics. Fourier transforms are introduced next and applied extensively to problems involving noise. Finally, two other transform methods are introduced and illustrated with examples.

In Chap. 1 the Laplace transform is introduced, and enough of the properties are derived to apply it to the solution of a simple differential equation. Details about convergence, uniqueness, and the like are discussed in Chap. 2. Transforms of higher derivatives and integrals and of some special functions are included here also.

The impulse function is the subject of Chap. 3. Its properties are derived, and it is then used to explain difficulties that occur at the origin when the Laplace transform is used. This chapter also includes a discussion of initial conditions.

Chapters 4 and 5 contain material showing the application of the Laplace-transform method to electrical networks. The equations are derived and illustrated with examples. Special topics include mutual inductance, switching operations, exchange of sources, and four-terminal networks.

In Chap. 6 the Laplace-transform method is applied to problems in dynamics. Lagrange's equations are used to write system equations, and the analyses of several simple gyroscope configurations are given as examples. Perturbation methods are discussed and applied to the motion of a satellite. Also included in this chapter is a discussion of analogs. The chapter concludes with the application of the Laplace-transform method to beam problems.

Inverse transforms are discussed in Chap. 7. Methods include the use of tables, partial fractions, residues, and poles and zeros. Various properties of the Laplace transform are discussed in Chap. 8. This is one of the most interesting and useful parts of the study of the transform method, and each property is illustrated with examples. Included here are discussions of nonconstant coefficient equations, partial differential equations, initial and final values, periodic functions, and many other special applications of the transform method.

Chapter 9 contains an account of the methods for describing linear systems. The Laplace transform furnishes a method for unifying these, which include the block diagram, impulse response, frequency response, transfer function, and s-plane geometry. Feedback systems are used in Chap. 10 to illustrate the application of the system function concept to

[1] Named for the Marquis Pierre-Simon de Laplace, who lived from 1749 to 1827, and whose interest was in applied mathematics.

analysis of a physical systems. Both root-locus and Nyquist methods are derived and illustrated with examples.

The partial differential equations describing the vibrating string, heat flow, and transient behavior of a transmission line are derived in Chap. 11. The Laplace transform is used to solve each for specific boundary conditions.

Chapter 12 introduces the Fourier series as a transform. The finite Fourier transform is applied to the solution of ordinary and partial differential equations. The Fourier integral is the subject of Chap. 13. Its application to system analysis is illustrated, and some of its properties are derived.

The next two chapters, 14 and 15, introduce the subject of noise in linear systems and the application of Fourier transforms to solution of noise problems. The emphasis is placed on the various ways that noise can be described. In Chap. 14 probability theory is discussed as the basis for methods of description derived from direct examination of random functions. The chapter ends with a discussion of the correlation function which leads to the definition of a frequency description in Chap. 15. There the spectral density of a noise signal is discussed, together with the problem of Fourier representation of a random process. Chapter 15 concludes with selected examples of the application of the theory to physical systems.

Chapter 16 contains a discussion of the solution of linear difference equations by a transform method. The transform appropriate to the problem is developed, and its application to physical problems illustrated.

The Mellin transform is introduced in Chap. 17 as an example of a method appropriate to the solution of a class of variable-coefficient equations. The chapter concludes with a summary of other transforms and their uses.

An introduction to the theory of functions of a complex variable has been included in Appendix A. This material leads to the derivation of the formulas for inversion of the Laplace transform and complex convolution. Appendix B contains various transform tables.

The material in this book has been used in several engineering courses at UCLA. Chapters 1 through 8 have been used in a course on transients at the senior and beginning graduate levels. Chapters 13 through 15 have been used as a part of a graduate course in advanced system analysis. A number of other combinations are possible: Replacing Chap. 6 by Chaps. 9 and 10, for example, places more emphasis on things electrical and on system analysis. Chapters 11 through 17 and Appendix A form a graduate-level course in the transform method.

I would like to thank many of my friends who have contributed in one way or another to the writing of this book. Professor John L. Barnes and

Prof. Louis Pipes of UCLA introduced me to engineering mathematics. I am especially grateful to Prof. Louis G. Walters, who used much of this material at UCLA and whose suggestions were most helpful. Dr. Arnold Rosenbloom and Prof. Harold Davis of UCLA were good enough to go over in detail the material on noise. My colleagues William E. Smith and Peter M. Kelly contributed helpful criticism throughout the writing. R. M. McClung, who supervised a correspondence course based on an earlier version of this book, contributed a number of helpful suggestions. I should also like to thank my students who were with me during the development of the material.

My special thanks go to my wife Jane, whose encouragement and patience made the writing of this book possible.

John A. Aseltine

CONTENTS

CHAPTER 1

THE LAPLACE TRANSFORM

1-1. Definition

We begin with a definition of the Laplace transformation upon which much of our linear system analysis is based:

The Laplace transformation is a mathematical operation indicated symbolically by

$$\mathcal{L}[f(t)]$$

[*spoken "Laplace transform of f(t)"*].

The operation indicated is[1]

$$\mathcal{L}[f(t)] \triangleq \int_0^\infty f(t)e^{-st}\,dt \triangleq F(s) \tag{1-1}$$

For the moment we accept this definition without going into it further. Later we shall discuss the conditions which must hold in order that the infinite integral exist. Most functions occurring in physical problems do have transforms, however, and the transform $F(s)$ is often simpler than the original function $f(t)$. The variable s is complex, and we shall have much more to say about this also.

1-2. Linear System

A linear system is one which can be described by a linear differential equation. Since all the differential equations we shall encounter in this book will be linear, we should say what a linear differential equation is and describe some of its properties.

Here is a differential equation:

$$P_3(x)\frac{d^3y}{dx^3} + P_2(x)\frac{d^2y}{dx^2} + P_1(x)\frac{dy}{dx} + P_0(x)y = f(x) \tag{1-2}$$

where $y = y(x)$ is called the dependent variable
$\quad x$ is called the independent variable[2]

---

[1] The symbol $\triangleq$ means "equal by definition."
[2] The letter used is not important. In many of our examples t will be the independent variable.

We say that a differential equation is linear when the dependent variable and its derivatives occur to the first degree only. This excludes powers, products, and functions such as sin y. The *order* of a differential equation is the same as the order of the highest derivative it contains. The general solution contains a number of arbitrary constants equal to the order.

Equation (1-2) is a linear differential equation of the third order, and its solution will contain three arbitrary constants. These constants are evaluated in physical problems by using information regarding the system at some special time or particular place in the system. If the time is zero, we call this information an *initial condition;* when a space dimension is involved, we call it a *boundary condition.*

A special kind of linear equation is one with constant coefficients:

$$A_3 \frac{d^3y}{dx^3} + A_2 \frac{d^2y}{dx^2} + A_1 \frac{dy}{dx} + A_0 y = f(x) \tag{1-3}$$

It is this kind of equation that the $\mathcal{L}$ transform can be applied to most effectively.

It is sometimes helpful in describing something to say what it is *not.* Here are examples of differential equations that are not linear:

$$\left(\frac{dy}{dt}\right)^2 + y = 1 \tag{1-4}$$

and
$$y \frac{d^2y}{dx^2} + \left(\frac{dy}{dx}\right)^2 = 0 \tag{1-5}$$

1-3. The Superposition Property

The property of superposition is characteristic of linear systems. Suppose that $y_1(x)$ is a solution of the linear equation

$$A_2(x) \frac{d^2y}{dx^2} + A_1(x) \frac{dy}{dx} + A_0(x)y = f_1(x) \tag{1-6}$$

That is,
$$A_2(x) \frac{d^2y_1}{dx^2} + A_1(x) \frac{dy_1}{dx} + A_0(x)y_1 \equiv f_1(x) \tag{1.7}$$

Furthermore, let $y_2(x)$ be the solution of

$$A_2(x) \frac{d^2y}{dx^2} + A_1(x) \frac{dy}{dx} + A_0(x)y = f_2(x) \tag{1-8}$$

That is,
$$A_2(x) \frac{d^2y_2}{dx^2} + A_1(x) \frac{dy_2}{dx} + A_0(x)y_2 \equiv f_2(x) \tag{1-9}$$

If we add Eqs. (1-7) and (1-9), we have

$$A_2(x) \frac{d^2}{dx^2} (y_1 + y_2) + A_1(x) \frac{d}{dx} (y_1 + y_2) + A_0(x)(y_1 + y_2)$$
$$\equiv f_1(x) + f_2(x) \quad (1\text{-}10)$$

In physical problems the function $f(x)$ is called the excitation or driving function; the function $y(x)$ is called the response. In words, Eq. (1-10) shows that if a system is linear, which is to say it has a linear differential equation, the response to a sum of inputs is simply the sum of the responses to these inputs applied separately. When we discuss physical systems, the reason for the terms *excitation* or *driving function* and *response* will be apparent.

The superposition property is an extremely useful one and can be taken as a defining property of linear systems. That is, a system for which superposition holds is a linear system.

1-4. Advantages of the Transform Method

The classical solution of an equation such as (1-3) is straightforward and can be found in any text on differential equations.[1] The transform method, however, offers advantages which make it well worth applying. Among these are the following:

1. The solution of the equation can usually be accomplished through the use of tables. In this respect the transform method resembles the method of multiplying numbers by adding their logarithms.

2. The initial conditions are carried along during the process of solution, so that there is no need to evaluate constants separately.

3. The transform method provides a great amount of intuition about linear physical systems.

These advantages will become apparent as we progress.

1-5. Transforms of Functions

The determination of the transform of a function follows from the definition (1-1):

$$\mathcal{L}[f(t)] \triangleq \int_0^\infty f(t)e^{-st} \, dt \triangleq F(s) \quad (1\text{-}11)$$

For example, the transform of e^{-at} is found as follows:

$$\mathcal{L}[e^{-at}] = \int_0^\infty e^{-at}e^{-st} \, dt = \int_0^\infty e^{-(a+s)t} \, dt$$
$$= \frac{-e^{-(a+s)t}}{a+s} \bigg|_0^\infty = \frac{1}{s+a} \quad (1\text{-}12)$$

---

[1] For example, see L. R. Ford, "Differential Equations," 2d ed., McGraw-Hill Book Company, Inc., New York, 1955.

The evaluation at the upper limit gives zero provided Re $[s] > -a$. Here Re means "real part of." If we write $s \triangleq \sigma + j\omega$ where $j = \sqrt{-1}$, this condition becomes $\sigma > -a$.

The result (1-12) can be tabulated:

$f(t)$	$F(s)$
e^{-at}	$\dfrac{1}{s+a}$

Another example: The unit step function $u(t)$ shown in Fig. 1-1:

$$\mathcal{L}[u(t)] = \int_0^\infty e^{-st}\, dt = -\left.\frac{e^{-st}}{s}\right|_0^\infty = \frac{1}{s} \tag{1-13}$$

We could continue to build up the table of transforms in this way. A short table is given below. On the left are functions in what we call the "time domain." On the right are the functions resulting from the application of the $\mathcal{L}$ operator (1-11). Since the variable in the right-hand column is s, we say that these entries are in the "s domain." Note that pair 6 reduces to pair 4 when we set $\alpha = 0$. It is often useful to set parameters equal to zero in this way to obtain a desired pair from a more general tabulated one.

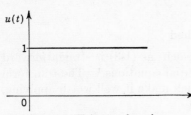

FIG. 1-1. Unit step function.

TABLE 1-1

$f(t)$	$F(s)$
1. $u(t)$	$\dfrac{1}{s}$
2. t	$\dfrac{1}{s^2}$
3. e^{-at}	$\dfrac{1}{s+a}$
4. $\sin \beta t$	$\dfrac{\beta}{s^2 + \beta^2}$
5. $\cos \beta t$	$\dfrac{s}{s^2 + \beta^2}$
6. $e^{-\alpha t} \sin \beta t$	$\dfrac{\beta}{(s + \alpha)^2 + \beta^2}$

1-6. Transforms of Operations

Another kind of transform pair concerns the result of applying the $\mathcal{L}$ operator to an operation in the time domain. Let us define

$$\mathcal{L}[f(t)] \triangleq F(s) \tag{1-14}$$

Now suppose we ask for

$$\mathcal{L}\left[\frac{df}{dt}\right] \triangleq \int_0^\infty \frac{df}{dt} e^{-st} dt \tag{1-15}$$

The presence of the derivative under the integral suggests integration by parts. Letting

$$u = e^{-st}$$
$$dv = \frac{df}{dt} dt \tag{1-16}$$

and using the formula

$$\int_a^b u \, dv = uv \Big|_a^b - \int_a^b v \, du \tag{1-17}$$

Eq. (1-15) becomes

$$\mathcal{L}\left[\frac{df}{dt}\right] = f(t)e^{-st} \Big|_0^\infty + s \int_0^\infty f(t)e^{-st} dt$$
$$= -f(0) + sF(s) \tag{1-18}$$

where $f(t)$ must be such that $\lim_{t \to \infty} f(t)e^{-\sigma t} = 0$ for some σ. As before, we can tabulate this result:

$f(t)$	$F(s)$
$\dfrac{df}{dt}$	$sF(s) - f(0)$

Another pair which follows from the definition of the transform is

$a_1 f_1(t) + a_2 f_2(t)$	$a_1 F_1(s) + a_2 F_2(s)$

$$\tag{1-19}$$

This one is basic in transforming differential equations.

1-7. Solution of a Differential Equation

We now have sufficient operation pairs and function pairs to solve a simple differential equation:

$$\frac{dy}{dt} + 4y = 2e^{-3t} \qquad y(0) = 3 \tag{1-20}$$

First, we transform both sides:

$$\mathcal{L}\left[\frac{dy}{dt} + 4y\right] = \mathcal{L}[2e^{-3t}] \tag{1-21}$$

Using pair (1;19),

$$\mathcal{L}\left[\frac{dy}{dt}\right] + 4\mathcal{L}[y] = 2\mathcal{L}[e^{-3t}] \tag{1-22}$$

Letting $Y(s) \triangleq \mathcal{L}[y(t)]$,

$$sY(s) - y(0) + 4Y(s) = \frac{2}{s+3}$$

$$Y(s)[s + 4] = \frac{2}{s+3} + 3$$

$$Y(s) = \frac{2}{(s+3)(s+4)} + \frac{3}{s+4}$$

$$= \frac{2}{s+3} + \frac{1}{s+4} \tag{1-23}$$

$Y(s)$ is the $\mathcal{L}$ transform of $y(t)$, the solution of the equation.[1]   Using the function-pair table, we find

$$y(t) = 2e^{-3t} + e^{-4t} \tag{1-24}$$

Notice that the initial condition is carried along throughout the solution.

1-8. Initial Condition

The number of initial conditions which must be provided for a complete solution is equal to the order of the equation.   In the example above, the initial value of the dependent variable was given.   Suppose, however, that $y'(0)$ were given instead.   For example,

$$\frac{dy}{dt} + 4y = 2e^{-3t} \qquad y'(0) = -2 \tag{1-25}$$

Transforming as before, we have

$$sY - y(0) + 4Y = \frac{2}{s+3} \tag{1-26}$$

and
$$Y = \frac{2}{(s+4)(s+3)} + \frac{y(0)}{s+4} \tag{1-27}$$

But $y(0)$ is not given.   To find $y(0)$ we write the differential equation for $t = 0$.   This can be done, since (1-25) must hold for any time:

$$y'(0) + 4y(0) = 2 \tag{1-28}$$

Substituting $y'(0) = -2$, we have

$$-2 + 4y(0) = 2$$
$$y(0) = 1 \tag{1-29}$$

---

[1] The final form (1-23) of $Y(s)$ is called a partial fraction and is especially convenient for finding inverse transforms from the tables.   This kind of expansion will be discussed in detail in Chap. 7.

Now the transform of the solution is

$$Y(s) = \frac{2}{(s+4)(s+3)} + \frac{1}{s+4} \tag{1-30}$$

1-9. Summary

We now have covered enough material so that if we had a larger table than Table 1-1 (and there is a larger one at the end of this book), we could use $\mathcal{L}$ transforms to solve differential equations. We have seen that the $\mathcal{L}$-transform method reduces the solving of linear, constant-coefficient differential equations to algebra and that the initial conditions are taken care of in the process. This is only the beginning, however. In the next chapter other properties of the $\mathcal{L}$ transform will be discussed, and in subsequent chapters we shall describe the use of the transform in studying and understanding the behavior of physical systems.

PROBLEMS

1-1. Which of the equations below are linear?

a. $2\frac{dy}{dt} + 3y^2 = 6t$

b. $t\frac{d^2y}{dt^2} + y = 0$

c. $\left(\frac{dy}{dt}\right)^2 + y = 3t^2$

d. $\frac{dy}{dx} + e^{-y} = x$

e. $\frac{dy}{dx} + e^{-x} = x$

1-2. Show that the superposition property does not hold for

$$\frac{dy}{dt} + y^2 = f(t)$$

1-3. Find

a. $\mathcal{L}[t]$
b. $\mathcal{L}[\sin \omega t]$
c. $\mathcal{L}[\cos \omega t]$
d. $\mathcal{L}[t^2]$

1-4. Confirm pair (1-19).

1-5. Solve for $y(t)$, and check by substituting the solution into the equation.

$$\frac{dy}{dt} + 2y = t \qquad y(0) = 0$$

1-6. From (1-30) write the solution to (1-25).

1-7. Solve for $i(t)$

$$L\frac{di}{dt} + Ri = E \sin \omega t \qquad i(0) = 0$$

1-8. Solve for $y(x)$

$$3\frac{dy}{dx} + y = e^{-x} \qquad y(0) = 1$$

1-9. Solve for $y(t)$, and check

$$\frac{dy}{dt} + 2y = \cos 3t \qquad y'(0) = 1$$

1-10. Show that if $v(t)$ has units of volts, then $\mathcal{L}[v(t)] = V(s)$ has units of volt-seconds.

CHAPTER 2

MORE ABOUT THE $\mathcal{L}$ TRANSFORM

2-1. Mathematical Preliminaries

In this book we emphasize the application of the $\mathcal{L}$ transform to the study and understanding of the behavior of physical systems. In general, the mathematical manipulations that we use will be formal. That is, we shall use a minimum amount of rigor. We should, however, state what some of the conditions are, and then we shall assume in the future that these hold for the problems considered.

Convergence. The integral which defines the Laplace transformation is an improper one, since one of its limits is infinity. This leads us to ask if there exists a finite transform for a given function. Without proof we state that a function $f(t)$ has a transform if the integral below exists:[1]

$$\int_0^\infty |f(t)| e^{-\sigma t}\, dt \tag{2-1}$$

That is, it must have a finite value for some real number σ. Most functions encountered in engineering satisfy this condition. This is because the exponential function $e^{-\sigma t}$ diminishes more rapidly than most other functions we encounter. For example, regardless of n,

$$\lim_{x \to \infty} x^n e^{-x} = 0 \tag{2-2}$$

An example of a function which does not have a $\mathcal{L}$ transform is $\exp(t^2)$. We can get around this difficulty for a physical system in the following way. Suppose we wish to apply e^{t^2} to a system and use $\mathcal{L}$ transforms to find the response. After some finite time t_0, the system will saturate if it contains vacuum tubes, motors, springs, etc. Operation in the saturated condition will lead to nonlinear equations at best, or destruction of equipment. The response in the linear range can be found by considering the input turned off before the system saturates:

$$\text{Input} = \begin{cases} e^{t^2} & t < t_0 \\ 0 & t \geq t_0 \end{cases} \tag{2-3}$$

This input meets the condition that (2-1) be finite.

[1] For a proof see D. V. Widder, "The Laplace Transform," p. 46, Princeton University Press, Princeton, N.J., 1946.

Uniqueness.   The $\mathcal{L}$ transform is defined as

$$\mathcal{L}[f(t)] \triangleq \int_0^\infty e^{-st}f(t)\,dt \triangleq F(s) \tag{2-4}$$

Since the integration starts at $t = 0$, the form of $f(t)$ for $t < 0$ cannot influence the transform.   For example, all the functions in Fig. 2-1 have the same $F(s)$.

In order that there be only one $f(t)$ for each $F(s)$, we shall say that a given $F(s)$ always corresponds to that time function which is zero for negative time [$f_1(t)$ in Fig. 2-1a].

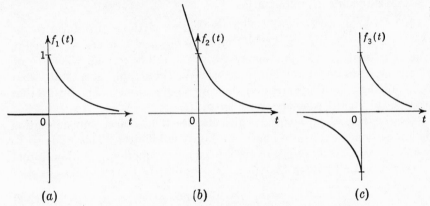

Fig. 2-1. Three functions which have the same $\mathcal{L}$ transform.

When we place this requirement on the time function for uniqueness, we imply the existence of a formula giving us $f(t)$ if we know $F(s)$.   The formula, which is derived in the Appendix, is

$$f(t) = \frac{1}{2\pi j} \int_{\sigma_1-j\infty}^{\sigma_1+j\infty} F(s)e^{st}\,ds \triangleq \mathcal{L}^{-1}[F(s)] \tag{2-5}$$

where σ_1 is greater than the σ of (2-1).   This is a contour integral and will not be discussed here.   We shall find inverse transforms by the use of tables built up from the direct transform.   This is analogous to the use of integral tables which are built up by taking derivatives.

Because we shall be dealing with functions which are discontinuous at the origin, like $f_1(t)$ in Fig. 2-1a, we need to say what we mean by $f_1(0)$. To avoid the discontinuity we shall approach $t = 0$ from the right.   All our initial values are to be interpreted in this way as a right-hand limit, written $f(0+)$.[1]   Since all our limits will be taken this way, we shall

---

[1] This is the same as saying that in the definition of the $\mathcal{L}$ transform we take the lower limit as $0+$: $\mathcal{L}[f(t)] \triangleq \lim_{\epsilon \to 0} \int_\epsilon^\infty f(t)e^{-st}\,dt$, $\epsilon > 0$.   More will be said about this definition in Sec. 3-6.

define

$$f(0) \triangleq f(0+) \tag{2-6}$$

in this book when we are solving differential equations by the $\mathcal{L}$ method. In the case of $f_1(t)$ in Fig. 2-1a, we have

$$f_1(0) \triangleq f_1(0+) = 1$$

2-2. Higher Derivatives

Let us consider first

$$\mathcal{L}[f''(t)] \triangleq \int_0^\infty f''(t)e^{-st}\,dt \tag{2-7}$$

We already have the operation pair (Sec. 1.6)

$$f'(t) \qquad \Big| \qquad sF(s) - f(0)$$

If we now define a new variable $g(t)$,

$$g(t) \triangleq f'(t) \tag{2-8}$$

we can write

$$\mathcal{L}[f''(t)] = \mathcal{L}[g'(t)] = sG(s) - g(0) \tag{2-9}$$

but

$$G(s) = \mathcal{L}[g(t)] = \mathcal{L}[f'(t)] = sF(s) - f(0) \tag{2-10}$$

and

$$g(0) = f'(0) \tag{2-11}$$

Now we can write

$$\mathcal{L}[f''(t)] = s^2F(s) - sf(0) - f'(0) \tag{2-12}$$

It is easy to see the pattern for the higher derivatives:

$$\mathcal{L}[f^{(n)}(t)] = s^nF(s) - s^{n-1}f(0) - s^{n-2}f'(0) - \cdots - f^{(n-1)}(0) \tag{2-13}$$

where

$$f^{(n)}(t) \triangleq \frac{d^nf}{dt^n} \tag{2-14}$$

Example. Let us apply the formula just obtained to a problem in the deflection of a beam. We shall go into this subject in some detail in Sec. 6-5. For the moment, we assume that if we have a beam as shown in Fig. 2-2, the equation describing the deflection under the influence of the loading shown is

$$EI\frac{d^4y}{dx^4} = f(x) \tag{2-15}$$

where $f(x)$ = force/unit length

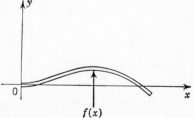

FIG. 2-2. A beam deformed by loading.

y = deflection

x = distance along beam

EI = a physical constant associated with material and shape of beam

We now transform (2-15) using the relation (2-13).

Letting

$$Y(s) \triangleq \mathcal{L}[y(x)]$$
$$F(s) \triangleq \mathcal{L}[f(x)] \qquad (2\text{-}16)$$

we have

$$s^4 Y(s) - s^3 y(0) - s^2 y'(0) - s y''(0) - y'''(0) = \frac{F(s)}{EI} \qquad (2\text{-}17)$$

The transform of the deflection $y(x)$ is then

$$Y(s) = \frac{y(0)}{s} + \frac{y'(0)}{s^2} + \frac{y''(0)}{s^3} + \frac{y'''(0)}{s^4} + \frac{F(s)}{s^4 EI} \qquad (2\text{-}18)$$

The "initial" conditions are various quantities evaluated at the origin. These are given names:

$$
\begin{aligned}
y &= \text{deflection} \\
y' &= \text{slope} \\
y'' &= (1/EI) \times \text{moment} \\
y''' &= (1/EI) \times \text{shear}
\end{aligned}
\qquad (2\text{-}19)
$$

and can be evaluated from a knowledge of the supports of the beam.

2-3. Integration

In the physical problems to come we shall need the $\mathcal{L}$ transform of the definite integral.[1]

$$\mathcal{L}\left[\int_0^t f(x)\, dx \right] = \int_0^\infty e^{-st}\, dt \int_0^t f(x)\, dx \qquad (2\text{-}20)$$

We integrate by parts, letting

$$
\begin{aligned}
u &= \int_0^t f(x)\, dx & dv &= e^{-st}\, dt \\
du &= f(t)\, dt & v &= -\frac{1}{s} e^{-st}
\end{aligned}
\qquad (2\text{-}21)
$$

Now

$$\int_0^\infty e^{-st}\, dt \int_0^t f(x)\, dx = -\frac{1}{s} e^{-st} \int_0^t f(x)\, dx \bigg|_0^\infty + \frac{1}{s} \int_0^\infty e^{-st} f(t)\, dt \qquad (2\text{-}22)$$

The first term on the right vanishes at the lower limit because $\int_0^0 = 0$.

[1] Note that the integral is a function only of its limits. Therefore,

$$\int_0^t f(t)\, dt = \int_0^t f(x)\, dx$$

We shall often use different letters for variables of integration.

At the upper limit, one of the properties[1] of functions which have transforms makes the term zero also. We now have

$$\mathcal{L}\left[\int_0^t f(x)\, dx\right] = \frac{F(s)}{s} \tag{2-23}$$

Example. The equation describing the electrical network shown in Fig. 2-3 is

$$Ri + \frac{1}{C}\int_0^t i\, dt = \frac{q(0)}{C} \tag{2-24}$$

where i = current

C = capacitance

R = resistance

$q(0)$ = initial charge on the capacitor

If we set $I(s) \triangleq \mathcal{L}[i(t)]$, we have for the transform of the equation after the switch closes

$$RI + \frac{I}{Cs} = \frac{q(0)}{Cs} \tag{2-25}$$

Solving for I,

$$I = \frac{q(0)}{Cs[R + (1/Cs)]} \tag{2-26}$$

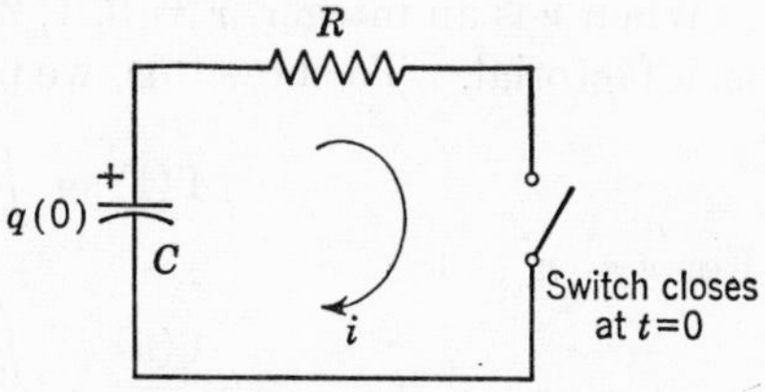

FIG. 2-3. Electrical network.

Note that the constant term on the right-hand side of (2-24) is treated as a step function. We shall discuss circuits at length in Chap. 4.

2-4. Transforms Involving Two Special Functions

In this section we shall illustrate some of the techniques which can be used to find transforms of functions.

Gamma Function. When we transform noninteger powers of t, an integral arises which cannot be expressed in terms of elementary functions. The problem of transforming t^ν will introduce this integral—called the gamma function—as well as some techniques which are useful in finding transforms of other functions.

To introduce the gamma function, let us transform t raised to the νth power

$$\mathcal{L}[t^\nu] = \int_0^\infty t^\nu e^{-st}\, dt \tag{2-27}$$

---

[1] If $\mathcal{L}[f(t)]$ exists for Re $[s] = \sigma > 0$, then $\lim_{t\to\infty} e^{-\sigma t}\int_0^t f(x)\, dx = 0$. Cf. Widder, *op. cit.*, p. 39.

We make the change of variable:

$$st = x \tag{2-28}$$

$$dt = \frac{dx}{s}$$

Then

$$\mathcal{L}[t^\nu] = \int_0^\infty \frac{x^\nu}{s^\nu} e^{-x} \frac{dx}{s} = \frac{1}{s^{\nu+1}} \int_0^\infty x^\nu e^{-x}\, dx \tag{2-29}$$

The integral on the right is a function only of ν, since x is the variable of integration. It is defined as

$$\Gamma(\nu + 1) \triangleq \int_0^\infty x^\nu e^{-x}\, dx \tag{2-30}$$

so that

$$\mathcal{L}[t^\nu] = \frac{\Gamma(\nu + 1)}{s^{\nu+1}} \tag{2-31}$$

When ν is an integer, $\nu = 0, 1, 2, \ldots$, the Γ function can be written as a factorial. To show this, we proceed as follows. For $\nu = 0$ we have

$$\Gamma(1) = \int_0^\infty e^{-x}\, dx = 1 \tag{2-32}$$

For $\nu = n - 1$

$$\Gamma(n) = \int_0^\infty x^{n-1} e^{-x}\, dx$$

If we integrate by parts letting

$$du = x^{n-1}\, dx \qquad u = \frac{1}{n} x^n \tag{2-33}$$

$$v = e^{-x} \qquad dv = -e^{-x}\, dx$$

we have

$$\Gamma(n) = \int_0^\infty x^{n-1} e^{-x}\, dx = \frac{1}{n} e^{-x} x^n \Big|_0^\infty + \frac{1}{n} \int_0^\infty x^n e^{-x}\, dx \tag{2-34}$$

The first term on the right vanishes at both limits. The integral on the right is $\Gamma(n + 1)$. Therefore,

$$\Gamma(n + 1) = n\Gamma(n) \tag{2-35}$$

Using (2-32) and (2-35) we can build up a table of values of $\Gamma(n)$ and values of $n! = n(n - 1) \ldots (2)(1) = n(n - 1)!$

$$\begin{aligned}
\Gamma(1) &= 1 \\
\Gamma(2) &= 1\Gamma(1) = 1 \cdot 1 = 1! \\
\Gamma(3) &= 2\Gamma(2) = 2 \cdot 1! = 2! \\
\Gamma(4) &= 3\Gamma(3) = 3 \cdot 2! = 3! \\
&\cdots\cdots\cdots\cdots\cdots \\
\end{aligned} \tag{2-36}$$

$$\Gamma(n + 1) = n\Gamma(n) = n(n - 1)! = n!$$

so we conclude that

$$\Gamma(n + 1) = n! \qquad (2\text{-}37)$$

The symbol 0! is equal to 1. This follows from setting $n = 1$ in the relation $n(n - 1)! = n!$. There-fore the relation (2-37) holds for $n = 0$.

When ν in Eq. (2-30) has values $-1, -2, \ldots$, it turns out that $\Gamma(\nu + 1) \to \infty$. A plot of $\Gamma(x)$ is shown in Fig. 2-4.

A special value which is some-times useful is[1]

$$\Gamma(\tfrac{1}{2}) = \sqrt{\pi} \qquad (2\text{-}38)$$

Error Function. Another func-tion which serves to illustrate some of the techniques useful in finding transforms is the Gauss curve.

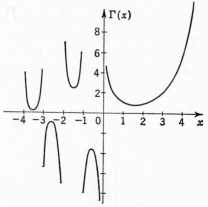

Fig. 2-4. Plot of the gamma function.

Many engineering problems, notably involving heat transfer and trans-mission lines, have solutions related to this curve. Later, we shall solve some of these problems by the £ transform. It will be useful to discuss the Gaussian function in order to be familiar with it when it occurs again.

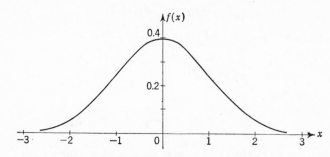

Fig. 2-5. Plot of Gauss curve with $\sigma = 1$.

The Gauss curve is shown in Fig. 2-5. It is described by

$$f(x) = \frac{1}{\sigma \sqrt{2\pi}} e^{-x^2/2\sigma^2} \qquad (2\text{-}39)$$

where σ is a constant called *standard deviation* and is a measure of the "spread" of the function.

[1] For a detailed account of the gamma function see, for example, L. A. Pipes, "Applied Mathematics for Engineers and Physicists," 2d ed., chap. 12, McGraw-Hill Book Company, Inc., New York, 1958.

A function which often occurs in the engineering problems is an integral of a function of this type:

$$\text{erf}\,(t) \triangleq \frac{2}{\sqrt{\pi}} \int_0^t e^{-x^2}\,dx \tag{2-40}$$

This integral cannot be expressed in terms of elementary functions and has been given the name *error function*, abbreviated erf (t). The error function has been tabulated in many places, just like the trigonometric functions.[1]   A graph of erf (t) is shown in Fig. 2-6.

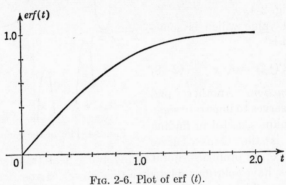

FIG. 2-6. Plot of erf (t).

Another related function is the *complementary error function*, defined as

$$\text{erfc}\,x \triangleq 1 - \text{erf}\,x \tag{2-41}$$

We note from Fig. 2-6 and relation (2-41) that

$$\begin{array}{ll}
\text{erf}\,(0) = 0 & \text{erfc}\,(0) = 1 \\
\text{erf}\,(\infty) = 1 & \text{erfc}\,(\infty) = 0
\end{array} \tag{2-42}$$

Let us now find

$$\mathcal{L}[\text{erf}\,(\sqrt{t})] = \mathcal{L}\left[\frac{2}{\sqrt{\pi}} \int_0^{\sqrt{t}} e^{-x^2}\,dx\right] \tag{2-43}$$

If we make the change of variable suggested by the square root in the upper limit

$$x^2 = y \tag{2-44}$$

then

$$dx = \frac{dy}{2\sqrt{y}} \tag{2-45}$$

The integral now has an upper limit t:

$$\int_0^{\sqrt{t}} e^{-x^2}\,dx = \frac{1}{2} \int_0^t \frac{e^{-y}}{\sqrt{y}}\,dy \tag{2-46}$$

---

[1] For example, H. B. Dwight, "Tables of Integrals and Other Mathematical Data," The Macmillan Company, New York, 1947, has erf $(t/\sqrt{2})$ tabulated on p. 237.

Now we can write

$$\mathcal{L}[\text{erf}\,(\sqrt{t})] = \mathcal{L}\left[\frac{1}{\sqrt{\pi}}\int_0^t \frac{e^{-y}}{\sqrt{y}}\,dy\right] \qquad (2\text{-}47)$$

But from our discussion of the transform of the definite integral in Sec. 2-3, we can write

$$\mathcal{L}\left[\frac{1}{\sqrt{\pi}}\int_0^t \frac{e^{-y}}{\sqrt{y}}\,dy\right] = \frac{1}{s}\mathcal{L}\left[\frac{1}{\sqrt{\pi}}\,\frac{e^{-t}}{\sqrt{t}}\right] \qquad (2\text{-}48)$$

Writing out the $\mathcal{L}$ operator, we have now

$$\mathcal{L}[\text{erf}\,(\sqrt{t})] = \frac{1}{\sqrt{\pi}}\,\frac{1}{s}\int_0^\infty \frac{e^{-st}e^{-t}}{\sqrt{t}}\,dt = \frac{1}{\sqrt{\pi}}\,\frac{1}{s}\int_0^\infty \frac{e^{-t(s+1)}}{\sqrt{t}}\,dt \quad (2\text{-}49)$$

We now make a second change of variable, letting $t(s+1) = x^2$; then

$$(s+1)\,dt = 2x\,dx \qquad (2\text{-}50)$$

$$dt = \frac{2\sqrt{t(s+1)}}{s+1}\,dx = \frac{2\sqrt{t}}{\sqrt{s+1}}\,dx \qquad (2\text{-}51)$$

We now have

$$\mathcal{L}[\text{erf}\,(\sqrt{t})] = \frac{1}{\sqrt{\pi}}\,\frac{1}{s}\int_0^\infty \frac{e^{-x^2}}{\sqrt{t}}\,\frac{2\sqrt{t}}{\sqrt{s+1}}\,dx$$

$$= \frac{2}{\sqrt{\pi}}\,\frac{1}{s}\,\frac{1}{\sqrt{s+1}}\int_0^\infty e^{-x^2}\,dx$$

$$= \frac{2}{\sqrt{\pi}}\,\frac{1}{s}\,\frac{1}{\sqrt{s+1}}\,\frac{\sqrt{\pi}}{2}\,\text{erf}\,(\infty) \quad (2\text{-}52)$$

So that

$$\mathcal{L}[\text{erf}\,(\sqrt{t})] = \frac{1}{s\sqrt{s+1}} \qquad (2\text{-}53)$$

It is worthwhile to follow the above derivation through carefully, not only because it makes use of the error function but because it illustrates two other common techniques. First, change of variable of integration is used. This is extremely useful in finding transforms. Second, a property of the transform, in this case relating to the definite integral, is used.

It is interesting to note that the $\mathcal{L}$ transforms of both t^ν and erf $(\sqrt{t})$ are irrational functions of s. These are the first transforms of this type we have encountered. We shall see in Chap. 11 that partial differential equations generally lead to irrational transforms.

2-5. Poles and Zeros

In much of our work with $\mathcal{L}$ transforms we shall treat s as a parameter. Actually as indicated in the inversion formula (2-5) s is a complex variable.

Values of s can be plotted in a complex s plane shown in Fig. 2-7. We shall be using this complex plane again, not in the actual solution of differential equations, but as a way of describing a function through its transform.   Before we can continue, we need the following definitions:

We say that $F(s)$ has a nth-order pole at $s = s_2$ if $\lim\limits_{s \to s_2} F(s) = \infty$ *and*

$[(s - s_2)^n F(s)]_{s=s_2}$ *is finite and nonzero.*

For example, if

$$F(s) = \frac{K(s + 2)}{s(s + 3)^2} \tag{2-54}$$

then $F(s)$ has a first-order pole (also called a "simple" pole) at $s = 0$ and a second-order pole at $s = -3$.

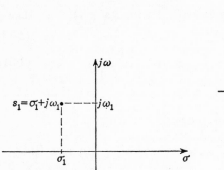

FIG. 2-7. Complex s plane.

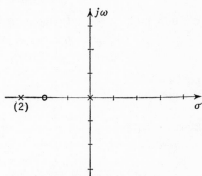

FIG. 2-8. Poles and zeros of $F(s) = K(s + 2)/s(s + 3)^2$.

We say that $F(s)$ has an nth-order zero at $s = s_3$ if $1/F(s)$ has an nth-order pole there.

In the example (2-54) $F(s)$ has a first-order zero at $s = -2$.

We can represent a function like (2-54) by a plot of its poles and zeros in the complex s plane.   Such a plot is shown in Fig. 2-8, where the zeros are marked (o) and the poles marked ($\times$).   Higher order poles are marked with the order number in parentheses.

The s-plane plot of poles and zeros can be used to reconstruct a rational $F(s)$,[1] for given Fig. 2-8, (2-54) could be written except for K. If we knew the value of $F(s)$ for some one value of s, K could be determined.

In Chap. 7 we shall show how the time function for a given $F(s)$ can be determined from a plot of the poles and zeros of $F(s)$.   To illustrate

---

[1] Poles and zeros are characteristic of rational functions.   Although

$$\mathcal{L}[t^\nu] = \frac{\Gamma(\nu + 1)}{s^{\nu+1}} \to \infty$$

as $s \to 0$, it has a pole at the origin only for integer values of ν.

TABLE 2-1. RELATION BETWEEN s-PLANE GEOMETRY AND CORRESPONDING TIME FUNCTION

$F(s)$	s plane	$f(t)$
$\dfrac{1}{s}$		$u(t)$
$\dfrac{1}{s^2}$		t
$\dfrac{1}{s + a}$		e^{-at}
$\dfrac{\beta}{s^2 + \beta^2}$		$\sin \beta t$
$\dfrac{s}{s^2 + \beta^2}$		$\cos \beta t$
$\dfrac{\beta}{(s + \alpha)^2 + \beta^2}$		$e^{-\alpha t} \sin \beta t$

the relationship between the s-plane geometry of $F(s)$ and the corresponding $f(t)$, the entries in the short Table 1-1 are shown in Table 2-1 with plots of the poles and zeros of $F(s)$.

The fact that a simple relationship exists between the time function and the geometry of its transform is very useful in system design. Considerable use is made of s-plane geometry in the study of feedback systems for example (see Chap. 10).

At the end of this book will be found a longer table of transforms and associated s-plane geometry.

PROBLEMS

2-1. Solve for $y(x)$:

$$\frac{d^2y}{dx^2} + 2\frac{dy}{dx} + y = u(x) \qquad y(0) = y'(0) = 0$$

2-2. Solve for $y(x)$:

$$\frac{d^2y}{dx^2} + 4y = \cos x \qquad y(0) = y'(0) = 0$$

2-3. Solve for $y(t)$:

$$\frac{d^2y}{dt^2} + 6\frac{dy}{dt} + 13y = 0 \qquad y(0) = 0 \qquad y'(0) = 2$$

2-4. Solve and check:

$$\frac{dy}{dt} + 3y + 2\int_0^t y\,dt = e^{-3t} - 2 \qquad y(0) = 2$$

2-5. Solve:

$$\frac{d^2y}{dt^2} + 4y = 0 \qquad \begin{matrix} y(0) = 0 \\ y'(0) = 1 \end{matrix}$$

2-6. Find

    a. $\Gamma(\tfrac{3}{2})$

    b. $\Gamma(-\tfrac{1}{2})$

    c. $\dfrac{\Gamma(n+1)}{\Gamma(n)}$

2-7. Show that

$$\mathcal{L}[e^{-at^2}] = \frac{1}{2}\sqrt{\frac{\pi}{a}}\,e^{s^2/4a}\,\text{erfc}\left(\frac{s}{2\sqrt{a}}\right)$$

2-8. Plot the poles and zeros of the functions:

    a. $\dfrac{1}{s^2/\omega_0^2 + (2\zeta/\omega_0)s + 1}$ when $\zeta = 0, 1$

    b. $\sin(s)$

2-9. Plot the locus of poles for Prob. 2-8*a* for

    a. $0 \leq \zeta \leq 1$ ω_0 constant

    b. $0 \leq \omega_0 \leq \infty$ ζ constant

2-10. What is the frequency of oscillation of the time function having a transform with poles as shown? *Ans.* $f = 1/2\pi$.

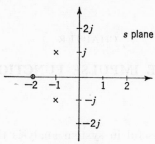

PROB. 2-10

2-11. Write $F(s)$ for the configurations shown:

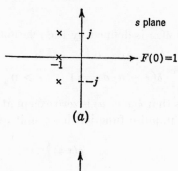

(a)

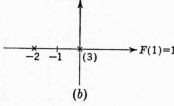

(b)

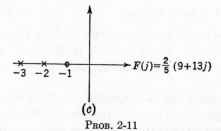

(c)

PROB. 2-11

CHAPTER 3

THE IMPULSE FUNCTION

One function is so useful in system analysis that it deserves special attention. This is the *impulse* or *delta function*. In this chapter we shall discuss the properties of the impulse function and show how it can be used in understanding the £-transform method.

3-1. Definition

The impulse function $\delta(x)$ is defined by the relations

$$\delta(x - a) = 0 \qquad x \neq a \tag{3-1}$$

$$\int_{a-\epsilon}^{a+\epsilon} \delta(x - a)\, dx = 1 \qquad \epsilon > 0 \tag{3-2}$$

Equation (3-1) indicates that $\delta(x - a)$ is zero except at $x = a$. Equation (3-2) requires that the impulse function have unit area.[1] Actually we

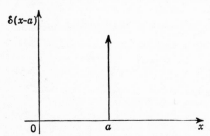

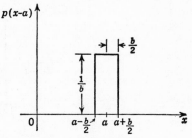

FIG. 3-1. Representation of a δ function.　　　FIG. 3-2. Square pulse.

call $\delta(x)$ a function rather loosely because the only place it has a nonzero value it is not defined.[2] The δ function is, however, very useful, as we shall see. If we wish to indicate this function on a plot, we do so as shown in Fig. 3-1.

The impulse function can be generated as a limit of a number of functions. For example, the Gaussian function of Sec. 2-4 will become a delta function if we let $\sigma \to 0$, since it has unit area required by (3-2)

---

[1] Because of the unit area $\delta(x)$ is sometimes called the *unit* impulse function.

[2] The rigorous mathematical basis for the impulse function is discussed by B. Friedman, "Principles and Techniques of Applied Mathematics," pp. 135–141, John Wiley & Sons, Inc., New York, 1956.

regardless of the value of σ. Another way to generate an impulse function is to begin with a square pulse with unit area as shown in Fig. 3-2. The unit area property holds regardless of the size of b, so that in the limit as $b \to 0$ the function becomes the impulse function.

3-2. Integrals

If an impulse function appears in an integrand multiplying another function, a particularly simple result is obtained. Let us consider the integral

$$\int_a^c f(x)\delta(x - b)\, dx \qquad a < b < c \qquad (3\text{-}3)$$

We note that the impulse occurs within the limits of integration. Now, since the impulse function is zero everywhere except at $x = b$, we can eliminate most of the range of integration:

$$\int_a^c f(x)\delta(x - b)\, dx = \int_{b-\epsilon}^{b+\epsilon} f(x)\delta(x - b)\, dx \qquad \epsilon > 0 \qquad (3\text{-}4)$$

and make ϵ as small as we like. Now if $f(x)$ is continuous at $x = b$, its change in value over the range of integration becomes very small as $\epsilon \to 0$. In fact, by making ϵ small we can make $f(x)$ over the range of integration as near to the constant $f(b)$ as we wish and write

$$\int_{b-\epsilon}^{b+\epsilon} f(x)\delta(x - b)\, dx \xrightarrow[\epsilon \to 0]{} f(b)\int_{b-\epsilon}^{b+\epsilon} \delta(x - b)\, dx \qquad \epsilon > 0 \qquad (3\text{-}5)$$

But the integral on the right is just the unit area of the impulse function regardless of the size of $\epsilon > 0$, so we finally write

$$\int_a^c f(x)\delta(x - b)\, dx = f(b) \qquad a < b < c \qquad (3\text{-}6)$$

One integral of special interest to us is the one with $f(x) = e^{-sx}$:

$$\mathcal{L}[\delta(x - \eta)] = \int_0^\infty e^{-sx}\delta(x - \eta)\, dx = e^{-s\eta} \qquad (3\text{-}7)$$

When we let η become small, we have

$$\lim_{\eta \to 0} \mathcal{L}[\delta(x - \eta)] = 1 \qquad \eta > 0 \qquad (3\text{-}8)$$

We must be careful to keep η positive by a small amount so that it will be within range of integration of the $\mathcal{L}$ integral (see Sec. 2-1), which does not include the origin.

Because of the simple $\mathcal{L}$ transform of the δ function, it will be useful in the description of physical systems. In Chap. 9, where we shall discuss the application of an impulse as an input to a system, the time of application will be a small time η later than $t = 0$ so that we can use (3-8). This delay will not make any difference physically but is necessary to keep the mathematics consistent.　Let us adopt the notation

$$\lim_{\eta \to 0} \delta(x - \eta) \triangleq \delta_+(x) \qquad \eta > 0 \tag{3-9}$$

where η is such that the impulse is included in the range of integration of the $\mathcal{L}$ integral.　We now have

$$\mathcal{L}[\delta_+(x)] = 1 \tag{3-10}$$

but
$$\mathcal{L}[\delta(x)] = 0 \tag{3-11}$$

This distinction may seem like splitting hairs, but its necessity will be brought out in the next section.

3-3. Relation to Step Function

Let us integrate the impulse function, leaving the upper limit variable:

$$u(t - a) = \int_0^t \delta(x - a)\, dx = \begin{cases} 0 & t < a \\ 1 & t > a \end{cases} \tag{3-12}$$

The function which we generate in this way can be plotted as in Fig. 3-3 and is recognized as the unit step function of Sec. 1-5.　We therefore can write

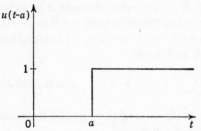

$$\frac{d}{dt} u(t - a) = \delta(t - a) \tag{3-13}$$

FIG. 3-3. Unit step function.

Here the d/dt symbol is used in the formal sense, since the limit implied by the derivative does not exist at a discontinuity.　We justify the use of this formal manipulation in the same way we justify the use of the impulse function as an aid in the understanding of physical systems.

Let us verify (3-13) by taking $\mathcal{L}$ transforms of both sides.　First we write [using the formula for the transform of a derivative (1-18)]

$$\mathcal{L}\left[\frac{d}{dt} u(t - a)\right] = s\mathcal{L}[u(t - a)] - u(-a) \tag{3-14}$$

Now $u(-a) = 0$ and

$$\begin{aligned}
\mathcal{L}[u(t - a)] &= \int_0^\infty u(t - a)e^{-st}\, dt \\
&= \int_a^\infty e^{-st}\, dt = \frac{e^{-as}}{s}
\end{aligned} \tag{3-15}$$

so that from (3-7)

$$\mathcal{L}\left[\frac{d}{dt} u(t - a)\right] = e^{-as} = \mathcal{L}[\delta(t - a)] \tag{3-16}$$

When we let $a \to 0$, the right side of (3-16) approaches $\mathcal{L}[\delta_+(t)]$ and (3-10) is verified. The formal application of the formula for the transform of the derivative of $u(t)$, on the other hand, gives

$$\mathcal{L}\left[\frac{d}{dt}\,u(t)\right] = \mathcal{L}[\delta(t)] = s\mathcal{L}[u(t)] - u(0+)$$

$$= s \cdot \frac{1}{s} - 1 = 0 \qquad (3\text{-}17)$$

which verifies (3-11) and shows the reason for introducing δ_+.

While we are on the subject of impulses and steps, it is worthwhile to talk about plotting these functions. It is helpful to think in terms of the *argument* of the functions [the argument of $f(x)$ is x]. We note from (3-1) that the impulse occurs where its argument is zero. From (3-12), the unit step function is zero for negative argument and unity for positive argument. To illustrate, we shall express the function in Fig. 3-2 in terms of steps. If we call the function in Fig. 3-2 $p(x - a)$, we can write it in several ways:

$$p(x - a) = \frac{1}{b}\left\{u\left[x - \left(a - \frac{b}{2}\right)\right] - u\left[x - \left(a + \frac{b}{2}\right)\right]\right\}$$

or

$$= \frac{1}{b}u\left[x - \left(a - \frac{b}{2}\right)\right]u\left[\left(a + \frac{b}{2}\right) - x\right] \qquad (3\text{-}18)$$

or

$$= \frac{1}{b}\left\{u\left[\left(a + \frac{b}{2}\right) - x\right] - u\left[\left(a - \frac{b}{2}\right) - x\right]\right\}$$

3-4. Derivatives

The derivatives of a function like $\delta(x)$ are highly improper, but still useful in many physical problems. Let us generate a δ function from a

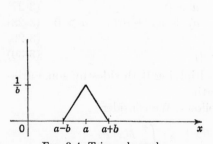

Fig. 3-4. Triangular pulse.

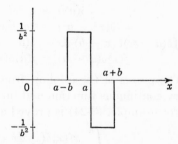

Fig. 3-5. Derivative of triangular pulse of Fig. 3-4.

triangular pulse, as in Fig. 3-4. The derivative of this function is shown in Fig. 3-5.

This function becomes the formal derivative of the impulse function when $b \to 0$ and is called the *unit doublet* and designated $\delta'(x)$.

As can be seen from Fig. 3-5, the integral of $\delta'(x)$ is zero. The moment is not zero, however:

$$\int_{a-\epsilon}^{a+\epsilon} x\delta'(x - a)\, dx = -1 \tag{3-19}$$

The doublet can be used in physical problems to represent, for example, a couple applied to a beam or the charge distribution of a dipole.

The Laplace transform of $\delta'(x - a)$ can be found by integrating by parts:

$$\int_0^\infty e^{-st}\delta'(t - a)\, dt = e^{-st}\delta(t - a)\,\Big|_0^\infty + \int_0^\infty se^{-st}\delta(t - a)\, dt \tag{3-20}$$

The first term on the right is zero because the δ function is zero at both limits. The integral on the right contains a δ function and can be evaluated by (3-6), so that

$$\mathcal{L}[\delta'(t - a)] = se^{-as} \tag{3-21}$$

Higher derivatives can be generated in the same way, and in general

$$\mathcal{L}[\delta^{(n)}(t - a)] = s^n e^{-as} \tag{3-22}$$

3-5. Other Properties of the Impulse Function

The impulse function has been used in formal manipulations extensively by Dirac.[1] Some of its properties are listed below. Dirac points out that these have meaning only when included as a part of an integrand, with a function which is continuous at the place where the impulse is located.

$$\delta(x) = \delta(-x) \tag{3-23}$$
$$x\delta(x) = 0 \tag{3-24}$$
$$\delta'(x) = -\delta'(-x) \tag{3-25}$$
$$x\delta'(x) = -\delta(x) \tag{3-26}$$
$$\delta(ax) = a^{-1}\delta(x) \qquad a > 0 \tag{3-27}$$
$$\delta(x^2 - a^2) = 2a^{-1}[\delta(x - a) + \delta(x + a)] \qquad a > 0 \tag{3-28}$$
$$\int \delta(a - x)\delta(x - b)\, dx = \delta(a - b) \tag{3-29}$$
$$f(x)\delta(x - a) = f(a)\delta(x - a) \tag{3-30}$$

The proofs of these follow from multiplying both sides by some arbitrary continuous function and integrating.

For example, (3-24) is proved as follows: We consider

$$\int_{-a}^{a} xf(x)\delta(x)\, dx = [xf(x)]_{x=0} \cdot \int_{-a}^{a} \delta(x)\, dx$$
$$= [xf(x)]_{x=0}$$
$$= 0 \tag{3-31}$$

provided that $f(x)$ is continuous at $x = 0$.

[1] P. A. M. Dirac, "The Principles of Quantum Mechanics," 2d ed., Oxford University Press, New York, 1935.

3-6. Troubles at the Origin

In Sec. 2-1, we said that initial conditions would be evaluated at $t = 0+$ in this book. Let us examine this choice a little more closely here, since we might have chosen $t = 0-$.

To fix ideas, we shall start with a very simple equation. Suppose we want to solve

$$\frac{dy}{dt} + y = 0 \qquad y(0) = 1 \qquad (3\text{-}32)$$

The solution is

$$y = e^{-t} \qquad (3\text{-}33)$$

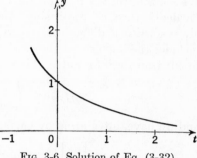

FIG. 3-6. Solution of Eq. (3-32).

and this is plotted in Fig. 3-6. We have said that when we use $\mathcal{L}$ transforms, we shall deal with functions which are zero for negative time. Let us call the solution to (3-32) which

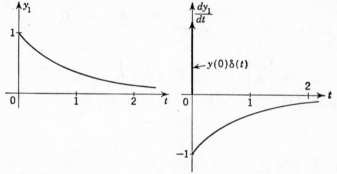

FIG. 3-7. Solution of (3-32) and its derivative valid for $t > 0$.

is valid for $t > 0$ but zero for $t < 0$:

$$y_1 = yu(t) \qquad (3\text{-}34)$$

The derivative of y_1 is

$$\frac{dy_1}{dt} = \frac{dy}{dt} u(t) + y\delta(t)$$

$$= \frac{dy}{dt} u(t) + y(0)\delta(t) \qquad (3\text{-}35)$$

This solution for $t > 0$ and its derivative are plotted in Fig. 3-7. We see that differentiation has introduced an impulse function at the origin. If we add the functions in (3-34) and (3-35), we obtain a new equation.

$$\frac{dy_1}{dt} + y_1 = y(0)\delta(t) + \left[y + \frac{dy}{dt}\right]u(t) \qquad (3\text{-}36)$$

The term in the brackets is zero according to (3-32), so that we can write

$$\frac{dy_1}{dt} + y_1 = y(0)\delta(t) \tag{3-37}$$

We see that for $t > 0$ this is the same as (3-32), since $\delta(t) = 0$ for $t > 0$. Evidently then, if we are using functions zero for negative time, we should always write a differential equation in a form such as (3-37), which includes the impulses at the origin. Let us now examine the lower limit in our transform formula. There are two possibilities. One is

$$\mathcal{L}_{0+}[f(t)] \triangleq \int_{\epsilon \to 0}^{\infty} f(t)e^{-st}\, dt \qquad \epsilon > 0 \tag{3-38}$$

and we have, integrating by parts,

$$\mathcal{L}_{0+}\left[\frac{df}{dt}\right] = [f(t)e^{-st}]_{\epsilon \to 0}^{\infty} + s \int_{\epsilon \to 0}^{\infty} f(t)e^{-st}\, dt$$
$$= sF(s) - f(0+) \tag{3-39}$$

The other way would be to define

$$\mathcal{L}_{0-}[f(t)] = \int_{-\epsilon \to 0}^{\infty} f(t)e^{-st}\, dt \qquad \epsilon > 0 \tag{3-40}$$

in which case

$$\mathcal{L}_{0-}\left[\frac{df}{dt}\right] = [f(t)e^{-st}]_{-\epsilon \to 0}^{\infty} + s \int_{-\epsilon \to 0}^{\infty} f(t)e^{-st}\, dt \tag{3-41}$$

Remembering that $f(t) = 0$ for $t < 0$, (3-41) becomes

$$\mathcal{L}_{0-}\left[\frac{df}{dt}\right] = sF(s) \tag{3-42}$$

Returning now to (3-37), we shall solve it using both forms. If we use $\mathcal{L}_{0+}$, the right-hand side is zero, since we do not include the origin in the transform definition. The transformed equation, using (3-39), is

$$sY(s) - y(0+) + Y(s) = 0 \tag{3-43}$$

If we use the $\mathcal{L}_{0-}$ form, we must include the origin, so that from (3-42) we have

$$sY(s) + Y(s) = y(0) \tag{3-44}$$

and we see that the same result is obtained either way.

We now conclude:

1. If we choose to use the $\mathcal{L}_{0-}$ form, we must always write a differential equation to include impulses at the origin. The equation

$$A_n \frac{d^n y}{dt^n} + \cdots + A_1 \frac{dy}{dt} + A_0 y = f(t)$$
$$\text{given } y(0), \ldots, y^{(n-1)}(0) \tag{3-45}$$

must be written

$$A_n \frac{d^n y}{dt^n} + \cdots + A_1 \frac{dy}{dt} + A_0 y = f(t) + A_n[y^{(n-1)}(0)\delta(t)$$
$$+ \cdots + y(0)\delta^{(n-1)}(t)] + \cdots + A_1 y(0)\delta(t) \quad (3\text{-}46)$$

2. If we use the $\mathcal{L}_{0+}$ form, we can omit the impulses at the origin from the equation. We must, however, be careful in taking the transform of an impulse near the origin [cf. (3-10) and (3-11)]:

$$\mathcal{L}_{0+}[\delta(t)] = 0$$
$$\mathcal{L}_{0+}[\delta_+(t)] = \mathcal{L}_{0+}[\delta(t-\eta)] = e^{-s\eta} \to 1 \quad \text{as } \eta \to 0 \quad (3\text{-}47)$$

Because it is somewhat more convenient to watch out for impulses near the origin than to modify every differential equation we want to solve, we shall use the $\mathcal{L}_{0+}$ form in this book. As can be seen from the discussion above, the choice is one of convenience.

3-7. Initial Conditions

Having adopted the $\mathcal{L}_{0+}$ form of the transform in the preceding section, we shall now show how initial conditions can be represented by impulse functions.

We have called the right-hand side of a differential equation which contains only the independent variable the "driving function." For example, in the equation

$$A_3 \frac{d^3 y}{dt^3} + A_2 \frac{d^2 y}{dt^2} + A_1 \frac{dy}{dt} + A_0 y = f(t) \quad (3\text{-}48)$$

$f(t)$ is the driving function. In an equation like (3-48), the complete solution requires three initial conditions: $y''(0)$, $y'(0)$, and $y(0)$. We shall show here that these initial conditions can be introduced as part of a new driving function.

Following the procedure of the last section for finding a solution of the equation which is zero for negative time, let us define

$$y_1(t) \triangleq y(t)u_+(t) \quad (3\text{-}49)$$

where $u_+(t)$ is a step beginning just to the right of the origin. We now write down the derivatives of $y_1(t)$:

$$\frac{dy_1}{dt} = \frac{dy}{dt} u_+(t) + y(t)\delta_+(t) \quad (3\text{-}50)$$

Now since $\delta_+(t)$ is zero everywhere except at $t = 0+$, we can write

$$\frac{dy_1}{dt} = \frac{dy}{dt} u_+(t) + y(0+)\delta_+(t) \quad (3\text{-}51)$$

Continuing,

$$\frac{d^2 y_1}{dt^2} = \frac{d^2 y}{dt^2} u_+(t) + \frac{dy(0+)}{dt} \delta_+(t) + y(0+)\delta'_+(t) \tag{3-52}$$

$$\frac{d^3 y_1}{dt^3} = \frac{d^3 y}{dt^3} u_+(t) + \frac{d^2 y(0+)}{dt^2} \delta_+(t) + \frac{dy(0+)}{dt} \delta'_+(t) + y(0+)\delta''_+(t) \tag{3-53}$$

Combining expressions for the derivatives, we write

$$
\begin{aligned}
A_3 \frac{d^3 y_1}{dt^3} + A_2 \frac{d^2 y_1}{dt^2} + A_1 \frac{dy_1}{dt} + A_0 y_1 = \Bigg[& A_3 \frac{d^3 y}{dt^3} + A_2 \frac{d^2 y}{dt^2} \\
& + A_1 \frac{dy}{dt} + A_0 y \Bigg] u_+(t) \\
& + [A_1 y(0+) + A_2 y'(0+) \\
& \qquad + A_3 y''(0+)]\delta_+(t) \\
& + [A_2 y(0+) + A_3 y'(0+)]\delta'_+(t) \\
& + [A_3 y(0+)]\delta''_+(t) \tag{3-54}
\end{aligned}
$$

Now from (3-48) the first term on the right can be written $f(t)u_+(t)$. Furthermore, from (3-49) the initial values of $y_1(t)$ and its derivatives are zero.

If we now adopt the convention of Sec. 2-1 and write $y(0)$ for $y(0+)$, we see that Eq. (3-48) can be solved as it stands with initial values of y specified, *or* we can solve

$$
\begin{aligned}
A_3 \frac{d^3 y_1}{dt^3} + A_2 \frac{d^2 y_1}{dt^2} + A_1 \frac{dy_1}{dt} + A_0 y_1 = & \; f(t)u_+(t) \\
& + [A_1 y(0) + A_2 y'(0) \\
& \qquad + A_3 y''(0)]\delta_+(t) \\
& + [A_2 y(0) + A_3 y'(0)]\delta'_+(t) \\
& + [A_3 y(0)]\delta''_+(t) \tag{3-55}
\end{aligned}
$$

with all initial conditions of y_1 equal to zero.   The solutions will be identical for $t > 0$, as specified in (3-49).   The difference in approach lies in the inclusion of initial conditions as part of the driving function in (3-55).

This technique can, of course, be generalized for linear equations of higher order than (3-48).   It will be useful later to be able to include initial conditions as a part of the driving function, especially when we discuss system functions (Sec. 9-1).

PROBLEMS

3-1. Confirm Eq. (3-19), using integration by parts.

3-2. Generate $\delta''(x - a)$ by considering a triangular approximation to $\delta'(x - a)$.

3-3. Show that

$$\int_{a-\epsilon}^{a+\epsilon} f(x)\delta'(x - a)\, dx = -f'(a)$$

3-4. Plot

 a. $u(-t)$

 b. $u(1 - t)$

 c. $tu(t - 1)$

 d. $u(t + a)u(2a - t)$

 e. $tu(t) - (t - 1)u(t - 1)$

3-5. Write an expression for each function below. Use step functions where necessary.

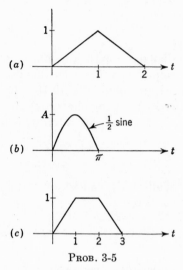

PROB. 3-5

3-6. Write the equations of Probs. 2-3, 2-4, and 2-5 with initial conditions replaced by impulses. Check by comparing transforms.

CHAPTER 4

ELECTRICAL NETWORKS

In this chapter we shall develop the differential equations for electrical networks. This development will be as direct as possible, since the reader will be to some degree familiar with the physical theory involved. Emphasis will be placed on getting the system equations ready for the $\mathcal{L}$-transform solution. We shall be interested particularly in initial conditions.

4-1. Preliminaries

The networks we shall study consist of certain elements connected together. These elements are

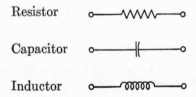

Resistor

Capacitor

Inductor

The equations describing the behavior of the network can be written in two ways. One way is called the loop method and has for dependent variables the currents flowing in the network. The other way is called the node method, the dependent variables being the voltages measured between points in the network.

The independent variable in network problems is time. There are two dependent variables: voltage and current. Sometimes the integral of the current, charge, is used. The symbols are

$$t = \text{time, sec (independent variable)}$$
$$v(t) = \text{voltage, volts (dependent variable)}$$
$$i(t) = \text{current, amp (dependent variable)}$$
$$q(t) = \text{charge, coulombs (dependent variable)}$$

Also

$$\frac{dq}{dt} \triangleq i \qquad q \triangleq \int_0^t i \, dt + q(0)$$

32

The relations between currents and voltages for the three kinds of elements are listed below. Also shown are sign conventions for positive voltage drops and the direction of positive assumed current flow.

1. Resistor

$$v = Ri \ (R \text{ in ohms}) \qquad i = \frac{1}{R} v \triangleq Gv$$

$$V(s) = RI(s) \qquad I(s) = GV(s)$$

2. Inductor

$$v = L \frac{di}{dt} \ (L \text{ in henrys}) \qquad i = \frac{1}{L} \int_0^t v \, dt + i(0)$$

$$V(s) = LsI(s) - Li(0) \qquad I(s) = \frac{1}{Ls} V(s) + \frac{i(0)}{s}$$

The initial current $i(0)$ is positive if it has the direction of i as shown. We sometimes write $1/L \triangleq \Gamma$.

3. Capacitor

$$v = \frac{1}{C} \int_0^t i \, dt + v(0) \ (C \text{ in farads}) \qquad i = C \frac{dv}{dt}$$

$$V(s) = \frac{1}{Cs} I(s) + \frac{v(0)}{s} \qquad I(s) = CsV(s) - Cv(0)$$

We sometimes write $1/C \triangleq S$, where S is called elastance. The initial voltage $v(0)$ is positive if it has the polarity of v as shown.

The currents and voltages which we measure in the network are the responses to excitations of voltage and current sources:

1. Voltage source

The automobile storage battery approximates an ideal voltage source.

2. Current source

$$i(t)$$

The pentode vacuum tube approximates an ideal current source.

Actually, neither of the examples cited above is a true ideal source, but only a good approximation. This is because the voltage and current measured at the terminals of the ideal sources above are defined to be independent of the network to which the source is connected. For example, the current source as shown above, being "ideal," causes a current $i(t)$ to flow across the terminals. Since there is an infinite resistance across the open-circuited terminals, there will be an infinite voltage drop across them. The "ideal" voltage source will cause a voltage $v(t)$ to appear across even a short circuit. These examples demonstrate the impossibility of physically realizing either of these "ideal" sources. The ideal sources are nevertheless useful approximations to real sources.

Since we shall use the $\mathfrak{L}$ method of solution on these systems, we must place a restriction on the sources that they be turned on at $t = 0$. This is because the functions of time used in the $\mathfrak{L}$ method are all assumed to be zero for $t < 0$ (see Sec. 2-1). Physically, this restriction is not a limitation because all systems are "turned on" at some given time. We call this "time zero" in our analysis.

4-2. Networks

An example of a network is shown in Fig. 4-1.

The *connections* in Fig. 4-1 can be shown symbolically by what is called a linear graph (Fig. 4-2). The terminology used in describing networks is shown in the figure.

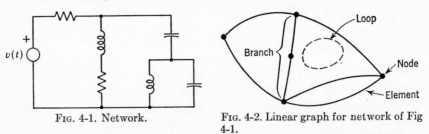

FIG. 4-1. Network.   FIG. 4-2. Linear graph for network of Fig 4-1.

The differential equations we write describing the behavior of a network are statements of Kirchhoff's laws for the particular network. There are two laws, so that each network can be described by two sets of equations, as we shall see. The laws state:

1. The sum of voltage drops in the elements of a loop is equal to the sum of applied voltages.

2. The sum of the currents leaving a node is equal to the sum of applied currents.

The choice of law for a given problem is arbitrary. Sometimes a simpler set of equations will result if one is chosen over the other, but either will provide a complete description of the network being studied.

In the writing of the equations, we shall use the following sign conventions:

1. Loop basis (Law 1): A positive current flows clockwise around the loop:

A positive applied voltage will have polarity as shown:

2. Node basis (Law 2): The voltage measured between each node and some single reference point (designated by the ground symbol ⏚) is positive.

A positive current source has polarity as shown:

If, after solving for the current or voltage, there results a negative quantity, that current or voltage will have direction or polarity opposite to that assumed.

In writing the equations for a network, it will be helpful to use matrix notation. Instead of writing the equations

$$a_{11}y_1(x) + a_{12}y_2(x) = f_1(x)$$
$$a_{21}y_1(x) + a_{22}y_2(x) = f_2(x)$$

(4-1)

we write

$$\begin{bmatrix} a_{11} & a_{12} \\ a_{21} & a_{22} \end{bmatrix} \begin{bmatrix} y_1(x) \\ y_2(x) \end{bmatrix} = \begin{bmatrix} f_1(x) \\ f_2(x) \end{bmatrix} \quad (4\text{-}2)$$

The bracketed terms in (4-2) are called matrices. We shall use this notation when we wish to bring out the properties of the coefficients of (4-1).

4-3. Loop Analysis

We illustrate the use of the definitions, laws, and conventions of the preceding section by applying them to a simple network. We shall then show how the transformed equations can be written down by inspection for a general network.

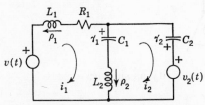

FIG. 4-3. Two-loop network.

Suppose we have the network shown in Fig. 4-3 with initial conditions as indicated. This is a two-loop network, so we shall have two equations for the unknown loop currents.

The condensers C_1 and C_2 have initial voltages at $t = 0$ of γ_1 and γ_2, respectively. The inductors L_1 and L_2 have initial currents ρ_1 and ρ_2, respectively.

Having assumed positive clockwise currents i_1 and i_2, we write the loop equations (Law 1) as follows.

For loop 1:

$$L_1 \frac{di_1}{dt} + R_1 i_1 + \frac{1}{C_1} \int_0^t (i_1 - i_2)\, dt + \gamma_1 + L_2 \frac{d}{dt}(i_1 - i_2) = v(t) \quad (4\text{-}3)$$

For loop 2:

$$\frac{1}{C_2} \int_0^t i_2\, dt - \gamma_2 + L_2 \frac{d}{dt}(i_2 - i_1) + \frac{1}{C_1} \int_0^t (i_2 - i_1)\, dt - \gamma_1$$
$$= -v_2(t) \quad (4\text{-}4)$$

Transforming, with $\mathcal{L}[i_1(t)] \triangleq I_1(s)$, etc., we have for loop 1

$$L_1 s I_1 + L_1 \rho_1 + R_1 I_1 + \frac{1}{C_1 s}(I_1 - I_2) + \frac{\gamma_1}{s} + L_2 s(I_1 - I_2) - L_2 \rho_2$$
$$= V(s) \quad (4\text{-}5)$$

and for loop 2:

$$\frac{1}{C_2 s} I_2 - \frac{\gamma_2}{s} + L_2 s(I_2 - I_1) + L_2 \rho_2 + \frac{1}{C_1 s}(I_2 - I_1) - \frac{\gamma_1}{s}$$
$$= -V_2(s) \quad (4\text{-}6)$$

Rearranging terms, we have

$$\left[(L_1 + L_2)s + R_1 + \frac{1}{C_1 s}\right] I_1 - \left[L_2 s + \frac{1}{C_1 s}\right] I_2$$

$$= V(s) - L_1\rho_1 - \frac{\gamma_1}{s} + L_2\rho_2 \quad (4\text{-}7)$$

$$-\left[L_2 s + \frac{1}{C_1 s}\right] I_1 + \left[L_2 s + \left(\frac{1}{C_2} + \frac{1}{C_1}\right)\frac{1}{s}\right] I_2$$

$$= -V_2(s) + \frac{\gamma_2}{s} + \frac{\gamma_1}{s} - L_2\rho_2 \quad (4\text{-}8)$$

These algebraic equations can be solved for I_1 and I_2.

Now we write the transformed equations in matrix form.

$$\begin{bmatrix} (L_1 + L_2)s + R_1 + \frac{1}{C_1 s} & -\left(L_2 s + \frac{1}{C_1 s}\right) \\ -\left(L_2 s + \frac{1}{C_1 s}\right) & L_2 s + \left(\frac{1}{C_2} + \frac{1}{C_1}\right)\frac{1}{s} \end{bmatrix} \begin{bmatrix} I_1 \\ I_2 \end{bmatrix}$$

$$= \begin{bmatrix} V(s) - L_1\rho_1 - \frac{\gamma_1}{s} + L_2\rho_2 \\ -V_2(s) + \frac{\gamma_2}{s} + \frac{\gamma_1}{s} - L_2\rho_2 \end{bmatrix} \quad (4\text{-}9)$$

We now make a number of observations about the equation above.

1. Let us write the matrix on the left as

$$\begin{bmatrix} z_{11}(s) & z_{12}(s) \\ z_{21}(s) & z_{22}(s) \end{bmatrix}$$

This is called the *impedance matrix*. We note the following:

a. $z_{11}(s)$ contains terms corresponding to all the elements in loop 1. $z_{22}(s)$ has terms for all the elements in loop 2. These are called the self-impedances of the loops. We shall show in Sec. 9-4 that setting $s = j\omega$ gives the impedance used in alternating-current theory.

b. $z_{12}(s)$ contains terms corresponding to the elements common to loops 1 and 2. We note that $z_{12} = z_{21}$. These are called mutual impedances. The terms in z_{12} are negative, whereas the terms in the self-impedances are positive.

2. The matrix on the right contains the sources or excitations. The initial conditions appear as sources. We can now write down rules for handling the initial conditions.

a. Initial voltage on capacitor can be treated as a constant voltage source in the loop.

b. Initial current in inductor appears in the transformed equation as a voltage source with transform $Li(0)$.

We note that the initial current, since it transforms to a constant, has the same effect as a source with voltage

$$v(t) = Li(0)\delta_+(t) \tag{4-10}$$

The sign will be positive if $i(0)$ is in the same direction as the loop current.

4-4. Loop Analysis—General Network

We can now generalize the results of the two-loop example above. The procedure for writing down the impedance matrix is as follows:

1. Draw a clockwise current in each loop, numbering each loop current.
2. Write down the z_{ij} terms as follows:
a. For $i \neq j$

$$z_{ij} = - \left(L_{ij}s + R_{ij} + \frac{S_{ij}}{s} \right) \tag{4-11}$$

where L_{ij} = total inductance common to loops i and j
R_{ij} = total resistance common to loops i and j
S_{ij} = total elastance common to loops i and j
(Note that if capacitors C_1, C_2, C_3 are common to loops i and j,

$$S_{ij} = \frac{1}{C_1} + \frac{1}{C_2} + \frac{1}{C_3} \tag{4-12}$$

b. For $i = j$

$$z_{jj} = L_{jj}s + R_{jj} + \frac{S_{jj}}{s} \tag{4-13}$$

where L_{jj} = total inductance in loop j
R_{jj} = total resistance in loop j
S_{jj} = total elastance in loop j

3. Write down the source term on the right-hand side of the equation, including for each loop:
a. All voltage sources.
b. Initially charged capacitors, treating them like constant voltage sources.
c. Terms due to initial currents in inductors [+ if $i(0)$ has same sense as loop current].

To illustrate the general loop procedure, we consider the network shown in Fig. 4-4 with initial conditions as indicated.

By inspection we write

$$\begin{bmatrix} s + 2 + \dfrac{2}{s} & -\left(2 + \dfrac{2}{s}\right) \\ -\left(2 + \dfrac{2}{s}\right) & s + 4 + \dfrac{4}{s} \end{bmatrix} \begin{bmatrix} I_1(s) \\ I_2(s) \end{bmatrix} = \begin{bmatrix} \dfrac{1}{s^2} - \dfrac{1}{s} \\ \dfrac{1}{s} - 1 \end{bmatrix} \tag{4-14}$$

The equations written out are

$$\left(s + 2 + \frac{2}{s}\right) I_1 - \left(2 + \frac{2}{s}\right) I_2 = \frac{1}{s^2} - \frac{1}{s}$$

$$-\left(2 + \frac{2}{s}\right) I_1 + \left(s + 4 + \frac{4}{s}\right) I_2 = \frac{1}{s} - 1 \tag{4-15}$$

Let us find the current $i_2(t) = \mathcal{L}^{-1}[I_2(s)]$:
Using Cramer's rule[1],

$$I_2 = \frac{\begin{vmatrix} s + 2 + \dfrac{2}{s} & \dfrac{1}{s^2} - \dfrac{1}{s} \\[2ex] -\left(2 + \dfrac{2}{s}\right) & \dfrac{1}{s} - 1 \end{vmatrix}}{\begin{vmatrix} s + 2 + \dfrac{2}{s} & -\left(2 + \dfrac{2}{s}\right) \\[2ex] -\left(2 + \dfrac{2}{s}\right) & s + 4 + \dfrac{4}{s} \end{vmatrix}} \tag{4-16}$$

[*Note:* Comparing this with (4-14) we see that the intermediate step of writing out the equations could have been avoided.]

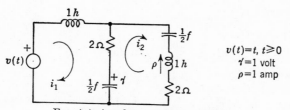

FIG. 4-4. Another two-loop network.

Now, expanding determinants, we have

$$\begin{aligned} I_2 &= \frac{(1/s^2)(s^2 + 2s + 2)(1 - s) + (1/s^3)(2s + 2)(1 - s)}{(1/s^2)(s^2 + 2s + 2)(s^2 + 4s + 4) - (1/s^2)(2s + 2)^2} \\[2ex] &= \frac{-s^4 - s^3 - 2s^2 + 2s + 2}{s(s^4 + 6s^3 + 10s^2 + 8s + 4)} \end{aligned} \tag{4-17}$$

We now must find the inverse transform of (4-17). This will be postponed until Chap. 7, where the method for handling a transform like (4-17) will be discussed.

[1] See, for example, E. A. Guillemin, "The Mathematics of Circuit Analysis," p. 15, John Wiley & Sons, Inc., New York, 1949.

4-5. Node Analysis

In loop analysis, the first of the Kirchhoff laws was used. We now illustrate the use of the second law which states that the sum of currents entering and leaving each node must be zero.

Suppose we have the network shown in Fig. 4-5.[1] Here the unknowns are the voltages v_1 and v_2 measured from the nodes to ground ($\perp$). Nodes for the equations are a little harder to find than loops. A good

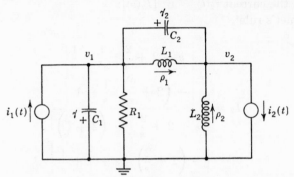

FIG. 4-5. Two-node network.

rule is to locate every point in the network where the voltage is unknown (the voltage at $\perp$ is assumed zero).

At node 1 we have

$$C_1 \frac{dv_1}{dt} + \frac{1}{R_1} v_1 + C_2 \frac{d}{dt}(v_1 - v_2) + \frac{1}{L_1} \int_0^t (v_1 - v_2)\, dt + \rho_1 = i_1 \quad (4\text{-}18)$$

and at node 2

$$\frac{1}{L_2} \int_0^t v_2\, dt - \rho_2 + C_2 \frac{d}{dt}(v_2 - v_1) + \frac{1}{L_1} \int_0^t (v_2 - v_1)\, dt$$
$$- \rho_1 = -i_2 \quad (4\text{-}19)$$

Transforming, these become (writing $1/R = G$)

$$C_1 s V_1 + C_1 \gamma + G_1 V_1 + C_2 s (V_1 - V_2) - C_2 \gamma_2 + \frac{1}{L_1 s}(V_1 - V_2)$$
$$+ \frac{\rho_1}{s} = I_1 \quad (4\text{-}20)$$

$$\frac{1}{L_2 s} V_2 - \frac{\rho_2}{s} + C_2 s(V_2 - V_1) + C_2 \gamma_2 + \frac{1}{L_1 s}(V_2 - V_1) - \frac{\rho_1}{s} = -I_2$$
$$(4\text{-}21)$$

[1] Strictly speaking the network of Fig. 4-5 should be called a two-*node-pair* network, since there are actually three nodes. We shall, however, classify networks according to the number of nodes at which the voltage is unknown—in this case two.

Upon rearranging terms, we have

$$\left[(C_1 + C_2)s + G_1 + \frac{1}{L_1 s}\right] V_1 - \left[C_2 s + \frac{1}{L_1 s}\right] V_2$$

$$= I_1 - C_1\gamma + C_2\gamma_2 - \frac{\rho_1}{s} \quad (4\text{-}22)$$

$$-\left[C_2 s + \frac{1}{L_1 s}\right] V_1 + \left[C_2 s + \left(\frac{1}{L_1} + \frac{1}{L_2}\right)\frac{1}{s}\right] V_2$$

$$= -I_2 + \frac{\rho_2}{s} - C_2\gamma_2 + \frac{\rho_1}{s} \quad (4\text{-}23)$$

In the matrix form this becomes

$$\left[\begin{array}{cc} (C_1 + C_2)s + G_1 + \dfrac{1}{L_1 s} & -\left(C_2 s + \dfrac{1}{L_1 s}\right) \\[2ex] -\left(C_2 s + \dfrac{1}{L_1 s}\right) & C_2 s + \left(\dfrac{1}{L_1} + \dfrac{1}{L_2}\right)\dfrac{1}{s} \end{array}\right] \left[\begin{array}{c} V_1 \\[2ex] V_2 \end{array}\right]$$

$$= \left[\begin{array}{c} I_1 - C_1\gamma + C_2\gamma_2 - \dfrac{\rho_1}{s} \\[2ex] -I_2 + \dfrac{\rho_2}{s} - C_2\gamma_2 + \dfrac{\rho_1}{s} \end{array}\right] \quad (4\text{-}24)$$

1. The matrix on the left of (4-24) is written

$$\left[\begin{array}{cc} y_{11} & y_{12} \\ y_{21} & y_{22} \end{array}\right]$$

and is called the *admittance matrix*. We note the following:

a. $y_{11}(s)$ contains terms corresponding to all elements connecting to node 1. $y_{22}(s)$ has terms for all elements connecting to node 2. These are called the self-admittances of the nodes.

b. $y_{12}(s)$ contains terms for all elements connecting nodes 1 and 2. Note that $y_{12} = y_{21}$. These are called mutual admittances. The terms in y_{12} are negative.

2. The source matrix on the right-hand side of (4-24) contains terms for the initial conditions as well as the actual current sources. Rules are:

a. Initial currents in inductors can be treated as constant-current sources.

b. Initial voltages on capacitors give rise to terms of the form $Cv(0)$. The same result could have been obtained from a current source

$$i(t) = Cv(0)\delta_+(t) \quad (4\text{-}25)$$

The sign will be positive if the polarity of $v(0)$ is such that the $+$ side of the capacitor connects to the node.

Before going on to the general method of node analysis, we comment that by now the reader probably has noticed a similarity with the loop method. This similarity will be formalized in Sec. 4-7 when we talk about duals.

4-6. Node Analysis—General Network

The transformed node equations for a network can be written down by inspection.

A general procedure is:

1. Label each node with a positive voltage. Mark a $\perp$ symbol on the reference node. (The reference node is chosen arbitrarily.)

2. Write down the y_{ij} terms in the admittance matrix as follows:

a. For $i \neq j$

$$y_{ij} = - \left(C_{ij}s + G_{ij} + \frac{\Gamma_{ij}}{s} \right) \tag{4-26}$$

where C_{ij} = total capacitance between nodes i and j
G_{ij} = total conductance between nodes i and j

$$\left(G_{\text{total}} = \frac{1}{R_1} + \frac{1}{R_2} + \cdots \right) \tag{4-27}$$

Γ_{ij} = total inverse inductance between nodes i and j

$$\left(\Gamma_{\text{total}} = \frac{1}{L_1} + \frac{1}{L_1} + \cdots \right) \tag{4-28}$$

b. For $i = j$

$$y_{jj} = C_{jj}s + G_{jj} + \frac{\Gamma_{jj}}{s} \tag{4-29}$$

where the C, G, and Γ are defined as totals connecting to the jth node.

3. Write the source matrix on the right, including for each node

a. All current sources.

b. Initial currents in inductors, treating them like constant current sources.

c. Terms due to initially charged capacitors.

Let us illustrate with the network shown in Fig. 4-6. We write by inspection

$$\begin{bmatrix} s + 2 + \dfrac{2}{s} & -\left(2 + \dfrac{2}{s}\right) \\ -\left(2 + \dfrac{2}{s}\right) & s + 4 + \dfrac{4}{s} \end{bmatrix} \begin{bmatrix} V_1(s) \\ V_2(s) \end{bmatrix} = \begin{bmatrix} \dfrac{1}{s^2} - \dfrac{1}{s} \\ \dfrac{1}{s} - 1 \end{bmatrix} \tag{4-30}$$

Equation (4-30) could be solved for the transform of $v_1(t)$ or $v_2(t)$.

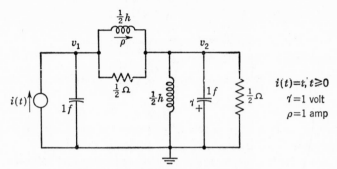

FIG. 4-6. Another two-node network.

4-7. Duals

A comparison of the discussions for the loop method and node method will reveal a striking similarity between the two. It seems that the two discussions are identical except for changes in words here and there. This similarity is not accidental but illustrates what is called the principle of *duality*. That is, the loop procedure is the same as the node procedure if certain changes, like interchanging the words "current" and "voltage," are made.

The similarity in the procedures suggests that there should be some similarity in the equations resulting. Comparing (4-9) and (4-24), we see that these two equations are of the same *form*. That is, if we changed the symbols around, the two equations would be identical. The networks described by these equations (Figs. 4-3 and 4-5) are called duals.

Definition. A network is the dual of another network if the equations describing one are of the same form as those describing the other, one set having current as dependent variable and the other having voltage.

It is possible to find the dual of a given network if the network can be drawn without wires crossing. Such a network is called *planar*. The procedure is as follows:

1. In each loop of the given network, place a dot. Each dot will be a node of the dual. Place one dot outside the network. This will be the reference node. Draw and number these dots on another diagram.

2. Draw lines on the original network between each dot in such a way that each element is crossed by a line. At the same time, draw corresponding lines on the other diagram, inserting dual elements:

inductor	—	capacitor
resistor	—	resistor
capacitor	—	inductor
voltage source	—	current source

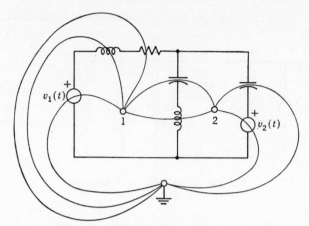

FIG. 4-7. Illustrating the method for finding a dual network.

The new diagram will be the dual of the first. To illustrate, let us show that Fig. 4-3 is the dual of Fig. 4-5 according to this graphical method. We already know they satisfy the definition of duals.

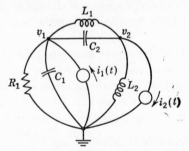

FIG. 4-8. Dual for network of Fig. 4-7.

We start with Fig. 4-7. Our other diagram is shown in Fig. 4-8. After a little straightening out, this will be identical with Fig. 4-5.

4-8. Loop Method vs. Node Method

A given network can be analyzed by either the loop method or the node method. Sometimes a saving in work can be achieved by using one or the other. For example, the network in Fig. 4-9 has three loops but only one unknown voltage v_1 and therefore one node. On the loop basis, three equations would be needed; on the node basis, only one. This suggests use of the node method. Sometimes there are more nodes than loops as in Fig. 4-10. This network has two nodes, one loop. Here the loop basis may reduce labor.

The method used must depend on the work involved and the variable desired. It should be pointed out that, since both of Kirchhoff's laws always hold, a combination can be used (see Example 2, Sec. 4-9).

We have presented just one way of choosing loops and nodes.   Others
are possible (for example, a loop might be chosen to include L_1, R_1, and

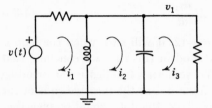

FIG. 4-9. Network with three loops and
one node.

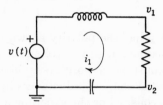

FIG. 4-10. Network with two nodes and
one loop.

C_2 in Fig. 4-3).   We shall not discuss here the problem of alternate choices
of loops and nodes.[1]

4-9. Mutual Inductance

When inductances are located near one another so that their magnetic
fields are linked, a changing current in one will induce voltages across the

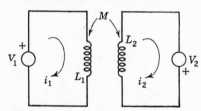

FIG. 4-11. Coupled inductances.

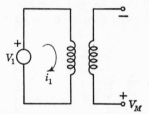

FIG.   4-12.   Experiment   to
determine sign of M.

others.   Let us consider the pair of coupled coils shown in Fig. 4-11.
We write the loop equations

$$L_1 \frac{di_1}{dt} \pm M \frac{di_2}{dt} = V_1 \tag{4-31}$$

$$\pm M \frac{di_1}{dt} + L_2 \frac{di_2}{dt} = V_2 \tag{4-32}$$

where we have included in each loop a voltage caused by changing current
in the other.   The voltage so induced is proportional to rate of change of
current.   The proportionality factor is called *mutual inductance*.   It may
be either positive or negative, depending on the physical arrangement of
the coils.   Let us perform an experiment to determine the sign of M.
Suppose we open the i_2 loop so that $i_2 = 0$ as shown in Fig. 4-12.   A

[1] For a complete discussion of network geometry, see E. A. Guillemin, "Introductory Circuit Theory," chap. 1, John Wiley & Sons, Inc., New York, 1953.

voltage V_M will appear across the terminals as shown due to the current i_1. This voltage is, from (4-32),

$$V_M = M\frac{di_1}{dt} \qquad (4\text{-}33)$$

If V_M has the polarity shown in Fig. 4-12, it will add, when the loop is closed again, to the voltage $L_2\,di_2/dt$, and in this case we conclude that M must be a positive number. We sometimes indicate the winding sense of coupled inductances by putting a dot at the positive ends of the two coils. For the experiment of Fig. 4-12, the dots would be placed as shown.

The sign of M may now be determined from the dots by noting that:

When both currents enter or both leave the ends of the inductances with dots, M is positive.

In Fig. 4-13a M is negative. In Fig. 4-13b M is positive.

When more than two inductances are coupled, the signs of the mutual inductances cannot always be determined from the positions of the dots.[1] We shall assume here, however, that when the dots are used, they are sufficient to determine sign.

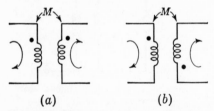
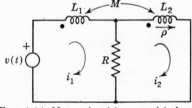

(a)   (b)

FIG. 4-13. M is a negative number in a and positive in b.   FIG. 4-14. Network with mutual inductance.

Example 1. Let us analyze the network shown in Fig. 4-14. We first write the equations

$$L_1\frac{di_1}{dt} + M\frac{di_2}{dt} + R(i_1 - i_2) = v$$
$$L_2\frac{di_2}{dt} + M\frac{di_1}{dt} + R(i_2 - i_1) = 0 \qquad (4\text{-}34)$$

Transforming,

$$L_1sI_1 + MsI_2 - M\rho + R(I_1 - I_2) = V$$
$$L_2sI_2 - L_2\rho + MsI_1 + R(I_2 - I_1) = 0 \qquad (4\text{-}35)$$

The matrix representation is

$$\begin{bmatrix} L_1s + R & Ms - R \\ Ms - R & L_2s + R \end{bmatrix}\begin{bmatrix} I_1(s) \\ I_2(s) \end{bmatrix} = \begin{bmatrix} V(s) + M\rho \\ L_2\rho \end{bmatrix}$$

[1] *Ibid.*, chap. 8.

Example 2: Combined Loop and Node Analysis. When a network contains mutual inductance, it is usually simpler to use loop analysis.

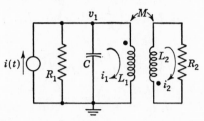

FIG. 4-15. Another network with mutual inductance.

However, it is possible to combine loop and node analysis in a single network. To illustrate, let us find the voltage v_1 in the network of Fig. 4-15. The node equation for v_1 is

$$\frac{v_1}{R_1} + C\frac{dv_1}{dt} + i_1 = i \tag{4-36}$$

Let us assume initial conditions are zero and transform:

$$\left(\frac{1}{R_1} + Cs\right)V_1 + I_1 = I \tag{4-37}$$

Now the loop equations for i_1 and i_2 are

$$L_1\frac{di_1}{dt} + M\frac{di_2}{dt} = v_1$$
$$M\frac{di_1}{dt} + L_2\frac{di_2}{dt} + R_2 i_2 = 0 \tag{4-38}$$

Transforming, we have

$$\begin{bmatrix} L_1 s & Ms \\ Ms & L_2 s + R_2 \end{bmatrix}\begin{bmatrix} I_1 \\ I_2 \end{bmatrix} = \begin{bmatrix} V_1 \\ 0 \end{bmatrix} \tag{4-39}$$

from which I_1 is found:

$$I_1 = \frac{\begin{vmatrix} V_1 & Ms \\ 0 & L_2 s + R_2 \end{vmatrix}}{\begin{vmatrix} L_1 s & Ms \\ Ms & L_2 s + R_2 \end{vmatrix}} = \frac{V_1(L_2 s + R_2)}{L_1 s(L_2 s + R_2) - M^2 s^2} \tag{4-40}$$

Substituting (4-40) into (4-37) we have

$$\left(\frac{1}{R_1} + Cs\right)V_1 + \frac{L_2 s + R_2}{L_1 s(L_2 s + R_2) - M^2 s^2}V_1 = I(s) \tag{4-41}$$

so finally

$$V_1 = \cfrac{L_1s(L_2s + R_2) - M^2s^2}{[(L_1L_2 - M^2)s^2 + L_1R_2s]\left(\dfrac{1}{R_1} + Cs\right) + (L_2s + R_2)} \, I(s)$$

$$= \cfrac{s\left(s + \dfrac{L_1R_2}{L_1L_2 - M^2}\right) I(s)}{C\left[s^3 + \left(\dfrac{1}{R_1C} + \dfrac{L_1R_2}{L_1L_2 - M^2}\right)s^2 + \left(\dfrac{L_1R_2}{L_1L_2 - M^2}\dfrac{1}{R_1C} + \dfrac{L_2}{(L_1L_2 - M^2)C}\right)s + \dfrac{R_2}{C(L_1L_2 - M^2)}\right]} \qquad (4\text{-}42)$$

PROBLEMS

4-1. Write in matrix form:

a. $\quad 2y_1 + 3y_2 = f_1$
$\quad\quad 4y_1 + 2y_2 = f_2$

b. $\quad i_1 - 2i_2 = t$
$\quad\quad -2i_1 + 3i_2 = 0$

c. $3x + 4y + 5z = 2$
$\quad 2x + 2y + 4z = 1$
$\quad 9x + 7y + z = 4$

4-2. What voltage appears across R_1 when $v(t)$ is a unit step?   Use the loop method.

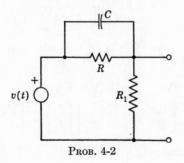

PROB. 4-2

4-3. Write the impedance matrix, and solve for $i_1(t)$.

$1\,h$

$v(t)$ $1\,\Omega$ i_1 i_2 $1\,h$ ρ $v(t) = e^{-t}$ $\rho = 1$ amp

PROB. 4-3

4-4. Write the impedance matrix equation for the network shown.

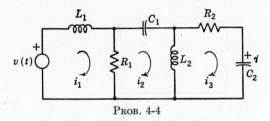

Prob. 4-4

4-5. Write the impedance matrix equation for the network shown.

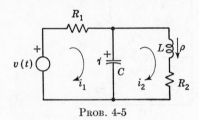

Prob. 4-5

4-6. Solve Prob. 4-2 by the node method.

4-7. Write admittance matrix equations for the networks shown.

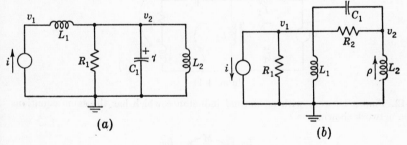

(a)　　　　　　　　　　　　(b)

Prob. 4-7

4-8. Solve for $v_1(t)$ by the node method.

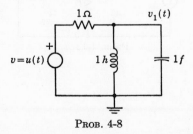

Prob. 4-8

4-9. Show that Figs. 4-4 and 4-6 are duals by the graphical method.

4-10. Find the duals of the networks in Probs. 4-7 and 4-8.

4-11. Write impedance matrix equations for the networks shown.

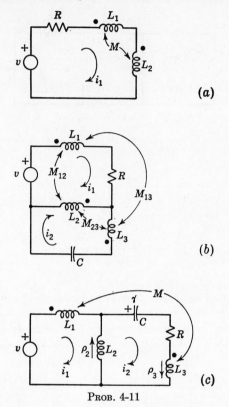

(a)

(b)

(c)

PROB. 4-11

4-12. Find a network with no mutual inductance which has the same equations as the network shown.

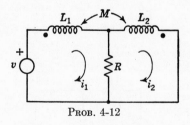

PROB. 4-12

CHAPTER 5

FURTHER TOPICS IN ELECTRICAL NETWORKS

In the last chapter we discussed methods for finding currents and voltages which resulted from the application of a disturbance to a general electrical network. Here we shall look into problems involving the opening and closing of switches in networks and discuss some applications of the transform method in network analysis. The applications include the concept of generalized impedance, exchange of current and voltage sources, and analysis of four-terminal networks.

5-1. Switching

In the electrical network problems of Chap. 4 the initial conditions were assumed known. In this section we shall go into the transient problem more deeply. All transients result from some change in a network or

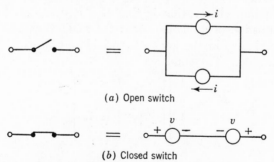

(a) Open switch

(b) Closed switch

Fig. 5-1. Equivalent sources for switches.

its sources, and this change can be brought about in general by opening or closing a switch. We shall now show how a switch can be replaced by an equivalent source, making possible the solution of switching problems by the methods of Chap. 4.

First, let us establish the equivalence of the switches and sources shown in Fig. 5-1. As an element, the *open* switch has zero current flowing between terminals, as does the pair of equal-current sources in Fig. 5-1a. The *closed* switch has zero voltage across terminals, as does the pair of equal-voltage sources in Fig. 5-1b.

Next we observe that if the voltage $v(t)$ exists across a pair of terminals, no current will flow from a voltage source of magnitude and polarity $v(t)$ placed across those terminals. The source will have no effect on the network. As an example,[1] consider the simple network shown in Fig.

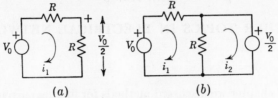

$$(a) \qquad (b)$$

Fig. 5-2. The additional voltage source in b leaves i_1 unchanged.

5-2a. The current i_2 in Fig. 5-2b is zero, and i_1 is the same for both networks. The equation for Fig. 5-2b is

$$\begin{bmatrix} 2R & -R \\ -R & R \end{bmatrix} \begin{bmatrix} i_1 \\ i_2 \end{bmatrix} = \begin{bmatrix} V_0 \\ -\dfrac{V_0}{2} \end{bmatrix}$$

$$i_2 = \frac{\begin{vmatrix} 2R & V_0 \\ -R & -\dfrac{V_0}{2} \end{vmatrix}}{\begin{vmatrix} 2R & -R \\ -R & R \end{vmatrix}} = 0 \tag{5-1}$$

A corresponding observation is that if a current $i(t)$ flows between nodes, the insertion of a current source of magnitude and polarity $i(t)$ will cause no change in the network variables. An example[2] of this is shown in Fig. 5-3a. The equation for Fig. 5-3b is

$$\begin{bmatrix} \dfrac{2}{R} & -\dfrac{1}{R} \\ -\dfrac{1}{R} & \dfrac{1}{R} \end{bmatrix} \begin{bmatrix} v_1 \\ v_2 \end{bmatrix} = \begin{bmatrix} i_0 \\ -\dfrac{i_0}{2} \end{bmatrix}$$

$$v_2 = \frac{\begin{vmatrix} \dfrac{2}{R} & i_0 \\ -\dfrac{1}{R} & -\dfrac{i_0}{2} \end{vmatrix}}{\begin{vmatrix} \dfrac{2}{R} & -\dfrac{1}{R} \\ -\dfrac{1}{R} & \dfrac{1}{R} \end{vmatrix}} = 0 \tag{5-2}$$

[1] The example is not a proof of the assertion that the additional source has no effect. The proof in the general case is omitted because of its complexity, but the pattern is the same as for our simple example.

[2] See the preceding footnote.

From these two observations—the equivalent circuits for switches and the possibility of adding sources without disturbance to a network—we can formulate a method for solving problems involving switches.

1. *Switch to be opened.* The switch is replaced by a current source equal to the current which would flow if the switch were left closed. At the instant of opening, a second equal and opposite-polarity current source is placed across the first. Using the superposition property (Sec. 1-3), the effect on any network variable is found by adding:

a. Effect due to original sources and initial conditions, with switch closed and the first current source (which changes nothing) in place.

b. Effect due to the second current source by itself with all other sources and initial conditions set equal to zero.

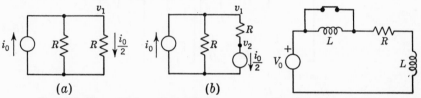

(a) (b)

FIG. 5-3. The voltage v_2 is zero in b.   FIG. 5-4. The switch is opened at $t = 0$.

2. *Switch to be closed.* In this case the open switch is replaced by a voltage source equal to the voltage which would appear across its terminals if it were left open. The closing is accomplished with a second source in series as in Fig. 5-1b. The superposition property is used as in the case just discussed.

Example 1. Suppose a network has been left as shown in Fig. 5-4 until all transients have died out. The switch is opened at $t = 0$.

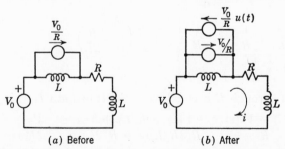

(a) Before   (b) After

FIG. 5-5. Before and after the switch of Fig. 5-4 is opened.

Before and after the opening of the switch the network is as shown in Fig. 5-5. The current i is the sum of two currents due to the sources shown in Fig. 5-6. From Fig. 5-6a,

$$i_0 = \frac{V_0}{R}$$

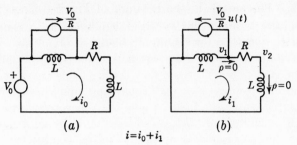

$$i = i_0 + i_1$$

FIG. 5-6. Components of current in Fig. 5-5b.

The current i_1 can be found by writing the transformed node equations for Fig. 5-6b:

$$\begin{bmatrix} \dfrac{1}{Ls} + \dfrac{1}{R} & -\dfrac{1}{R} \\[2mm] -\dfrac{1}{R} & \dfrac{1}{Ls} + \dfrac{1}{R} \end{bmatrix} \begin{bmatrix} V_1 \\[2mm] V_2 \end{bmatrix} = \begin{bmatrix} -\dfrac{V_0}{Rs} \\[2mm] 0 \end{bmatrix} \tag{5-3}$$

$$V_1 = \frac{\begin{vmatrix} -\dfrac{V_0}{Rs} & -\dfrac{1}{R} \\[2mm] 0 & \dfrac{1}{Ls} + \dfrac{1}{R} \end{vmatrix}}{\begin{vmatrix} \dfrac{1}{Ls} + \dfrac{1}{R} & -\dfrac{1}{R} \\[2mm] -\dfrac{1}{R} & \dfrac{1}{Ls} + \dfrac{1}{R} \end{vmatrix}} = \frac{-\dfrac{V_0}{R^2 s^2}\left(s + \dfrac{R}{L}\right)}{\dfrac{\left(s + \dfrac{R}{L}\right)^2}{R^2 s^2} - \dfrac{1}{R^2}} = \frac{-V_0\left(s + \dfrac{R}{L}\right)}{2\dfrac{R}{L}\left(s + \dfrac{R}{2L}\right)} \tag{5-4}$$

$$V_2 = \frac{\begin{vmatrix} \dfrac{1}{Ls} + \dfrac{1}{R} & -\dfrac{V_0}{Rs} \\[2mm] -\dfrac{1}{R} & 0 \end{vmatrix}}{\begin{vmatrix} \dfrac{1}{Ls} + \dfrac{1}{R} & -\dfrac{1}{R} \\[2mm] -\dfrac{1}{R} & \dfrac{1}{Ls} + \dfrac{1}{R} \end{vmatrix}} = \frac{-V_0 s}{2\dfrac{R}{L}\left(s + \dfrac{R}{2L}\right)}$$

The current through R is equal to i_1; the transform $I_1(s)$ is

$$I_1(s) = \frac{V_1(s) - V_2(s)}{R} = \frac{-V_0(s + R/L) + V_0 s}{R(2R/L)(s + R/2L)} = -\frac{V_0}{2R}\frac{1}{s + R/2L} \tag{5-5}$$

Then

$$I = I_0 + I_1 = \frac{V_0}{R}\left[\frac{1}{s} - \frac{1}{2}\frac{1}{s + (R/2L)}\right]$$

$$i(t) = \frac{V_0}{R}\left[1 - \frac{1}{2}e^{-(R/2L)t}\right] \tag{5-6}$$

The current is plotted in Fig. 5-7.

Example 2. The switch in the network of Fig. 5-8 closes at $t = 0$. Before $t = 0$, the network has come to equilibrium. The conditions before and after closing of the switch are shown in Fig. 5-9. Let us find the current i in the first loop after the switch closes. This current will

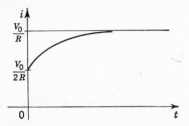

FIG. 5-7. Current in network of Fig. 5-4.

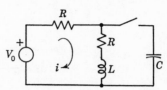

FIG. 5-8. The switch closes at $t = 0$.

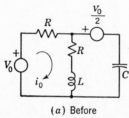

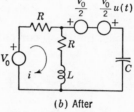

(*a*) Before   (*b*) After

FIG. 5-9. Before and after the switch of Fig. 5-8 is closed.

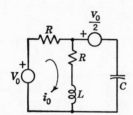

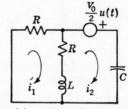

FIG. 5-10. Current in Fig. 5-9*b* is the superposition of i_0 and i_1 in these networks.

be the sum of the currents flowing in the corresponding loops of the networks of Fig. 5-10 which are derived from Fig. 5-9*b*. The desired current is

$$i = i_0 + i_1 \qquad (5\text{-}7)$$

The steady current flowing before the switch is closed is

$$i_0 = \frac{V_0}{2R} \qquad (5\text{-}8)$$

The transformed loop equations for Fig. 5-10*b* are

$$\begin{bmatrix} Ls + 2R & -(Ls + R) \\ -(Ls + R) & Ls + R + \dfrac{1}{Cs} \end{bmatrix} \begin{bmatrix} I_1 \\ I_2 \end{bmatrix} = \begin{bmatrix} 0 \\ \dfrac{V_0}{2s} \end{bmatrix} \qquad (5\text{-}9)$$

from which

$$I_1 = \frac{\begin{vmatrix} 0 & -(Ls + R) \\ \dfrac{V_0}{2s} & Ls + R + \dfrac{1}{Cs} \end{vmatrix}}{\begin{vmatrix} Ls + 2R & -(Ls + R) \\ -(Ls + R) & Ls + R + \dfrac{1}{Cs} \end{vmatrix}}$$

$$= \frac{\dfrac{V_0}{2s}(Ls + R)}{(Ls + 2R)\left(Ls + R + \dfrac{1}{Cs}\right) - (Ls + R)^2}$$

$$= \frac{V_0}{2R}\frac{s + \dfrac{R}{L}}{s^2 + \left(\dfrac{1}{RC} + \dfrac{R}{L}\right)s + \dfrac{2}{LC}} \qquad (5\text{-}10)$$

The transformed first-loop current is finally

$$I = I_0 + I_1 = \frac{V_0}{2Rs} + \frac{V_0}{2R}\frac{s + \dfrac{R}{L}}{s^2 + \left(\dfrac{1}{RC} + \dfrac{R}{L}\right)s + \dfrac{2}{LC}}$$

$$= \frac{V_0}{R}\frac{s^2 + \left(\dfrac{1}{2RC} + \dfrac{R}{L}\right)s + \dfrac{1}{LC}}{s\left[s^2 + \left(\dfrac{1}{RC} + \dfrac{R}{L}\right)s + \dfrac{2}{LC}\right]} \qquad (5\text{-}11)$$

In these examples the networks were assumed to be in equilibrium before switching. The same technique can be used to solve problems in which several successive switching operations take place.

5-2. Impedance

Considerable time is usually devoted to the topic of impedance in the study of alternating current. A certain amount of mystery usually surrounds the subject for students because impedance is, in general, a complex number.

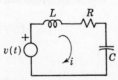

FIG. 5-11. Simple LRC network.

Now, with some $\mathcal{L}$-transform background, a more general impedance can be defined in a straightforward way. The connection with a-c circuit theory will be mentioned here, but details will be postponed until Chap. 9.

A simple LRC network is shown in Fig. 5-11. Let us assume that all initial conditions are included in $v(t)$ as discussed in Sec. 3-7. Then the

equation is

$$L\frac{di}{dt} + Ri + \frac{1}{C}\int_0^t i\, dt = v(t) \tag{5-12}$$

and transforming we have

$$\left(Ls + R + \frac{1}{Cs}\right)I(s) = V(s) \tag{5-13}$$

We now take the ratio of the transforms of voltage to current:

$$\frac{V(s)}{I(s)} = Ls + R + \frac{1}{Cs} \tag{5-14}$$

For networks composed of L, R, and C elements as we have defined them (Sec. 4-1), this ratio will always be a function only of the element values; it is independent of the form of driving function. We call this ratio the *generalized impedance* of the network and use the symbol $Z(s)$.

$$Z(s) \triangleq \frac{V(s)}{I(s)} \tag{5-15}$$

We can measure an impedance at any pair of nodes of a general network. The usual way of representing this is to show two terminals connected to the nodes protruding from a "black box" containing the network as in Fig. 5-12. The black box can contain any number of Rs, Ls, and Cs connected in any fashion. It must, however, contain no voltage or current sources.

Fig. 5-12. Black box.

The reciprocal of $Z(s)$, called *admittance*, is often used:

$$Y(s) \triangleq \frac{1}{Z(s)} \tag{5-16}$$

The impedance function was first used in the study of the steady-state[1] behavior of networks with sinusoidal signals applied. The impedance function defined in (5-15) is much more general, since $V(s)$ and $I(s)$ need not be transforms of sinusoids. The two impedances are connected very simply: The sinusoidal impedance is $Z(j\omega)$, where s has been replaced by $j\omega$. Here j is the square root of (-1) and ω is the frequency of the sinusoid in radians per second.

If a voltage $V(t) = \sin \omega t$ is applied to the black box in Fig. 5-12, the current flowing in the steady state at the terminals will be

$$i(t) = |Y(j\omega)| \sin(\omega t + \phi) \tag{5-17}$$

[1] The steady-state condition is reached when all transients caused by the application of the input have died out.

where ϕ is the angle of the complex number $Y(j\omega)$. The proof for this will be presented in Chap. 9.

From (5-15), if a voltage source with transformed voltage $V(s)$ is applied to the black box of Fig. 5-12, then the transform of the current flowing at the terminals will be

$$I(s) = V(s)Y(s) \tag{5-18}$$

If a current source is applied, the voltage across the terminals will be

$$V(s) = I(s)Z(s) \tag{5-19}$$

We see that the impedance is a way of describing a network because it can be used to find the response to an arbitrary input applied at $t = 0$ as well as the steady-state response to a sinusoid. We shall discuss this means of describing a physical system in detail in Chap. 9.

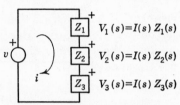

FIG. 5-13. Series connection.

When we have several networks, each with its own impedance, the impedance of a combination of these can be found in a straightforward way. For the series arrangement shown in Fig. 5-13, the same current must flow in each black box. The voltage applied must be equal to the sum of voltages in the loop. Hence

$$\begin{aligned} V(s) &= V_1(s) + V_2(s) + V_3(s) \\ &= I(s)[Z_1(s) + Z_2(s) + Z_3(s)] \end{aligned} \tag{5-20}$$

The impedance of the combination is $V(s)/I(s)$ so that

$$Z = Z_1 + Z_2 + Z_3 \tag{5-21}$$

For the parallel connection shown in Fig. 5-14, the same voltage is

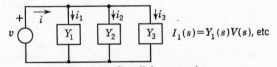

FIG. 5-14. Parallel connection.

applied to each black box. The current i is the sum of currents flowing in the branches:

$$\begin{aligned} I(s) &= I_1(s) + I_2(s) + I_3(s) \\ &= V(s)[Y_1(s) + Y_2(s) + Y_3(s)] \end{aligned} \tag{5-22}$$

The admittance of the combination is $I(s)/V(s)$, so that

$$Y = Y_1 + Y_2 + Y_3 \tag{5-23}$$

5-3. Exchange of Sources

We can use the impedance concept to derive means for exchanging voltage and current sources in a network. This exchange is often useful in simplifying analysis.

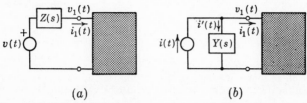

(a)   (b)

FIG. 5-15. Voltage and current sources applied to a black box.

Our aim will be to find conditions for equivalence between the two network configurations shown in Fig. 5-15. The two black boxes are identical.

Referring to the Fig. 5-15a we have the following relation:

$$V_1(s) = V(s) - I_1(s)Z(s) \tag{5-24}$$

In Fig. 5-15b we have

$$I_1(s) = I(s) - I'(s) \tag{5-25}$$
$$= I(s) - V_1(s)Y(s)$$

or
$$V_1(s) = \frac{I(s)}{Y(s)} - \frac{I_1(s)}{Y(s)} \tag{5-26}$$

Now any measurement we might make inside the black box would remain unchanged as long as v_1 and i_1 do not change. Comparing (5-24) and (5-26) we see that internal measurements will be identical for Fig. 5-15a or b if

$$I(s) = V(s)Y(s)$$
$$Y(s) = \frac{1}{Z(s)} \tag{5-27}$$

These relations (5-27) enable us to replace a voltage source and series impedance by a current source and shunt admittance, and vice versa. The examples that follow illustrate the use of exchange of sources in network analysis.

Example 1. For small signals a vacuum tube can be represented as shown in Fig. 5-16. The parameters μ and r_p are characteristics of the tube, called *amplification factor* and *plate resistance*, respectively. The voltage e_g is applied to the grid; e_p is the plate voltage. The resistance R_L is called the *load*.

The voltage source can be replaced by a current source by using (5-27). The equivalent circuit (identical voltage across the load) is shown in Fig. 5-17. The ratio μ/r_p is called *transconductance* and is given the symbol g_m.

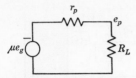

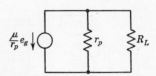

FIG. 5-16. Voltage-source equivalent circuit of a vacuum tube.

FIG. 5-17. Current-source equivalent cuit of a vacuum tube.

Since both circuits cause the same current to flow in the load, either can be used. However, Fig. 5-16 is most useful in the analysis of triodes where r_p is small; Fig. 5-17 is used for pentodes where r_p is very large.

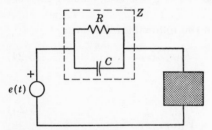

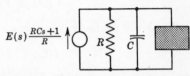

FIG. 5-18. Network with series impedance.

FIG. 5-19. Equivalent network for Fig. 5-18.

Example 2. The relations (5-27) hold in general, and the impedance need not be a resistance as in the previous example. To illustrate, suppose we start with the network shown in Fig. 5-18. In this case

$$Z = \frac{1}{(1/R) + Cs} = \frac{R}{RCs + 1} \tag{5-28}$$

If we now label the source with its $\mathcal{L}$ transform, an equivalent network, as far as the black box is concerned, is shown in Fig. 5-19. If the R and

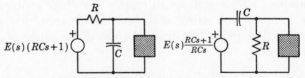

FIG. 5-20. More equivalent circuits.

the C are considered separately as the shunt to the current source, two more equivalent circuits can be derived as shown in Fig. 5-20. Finally the sources can be labeled with the time functions as shown in Fig. 5-21

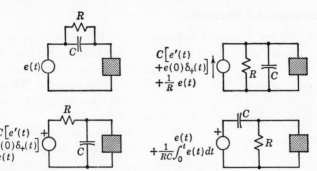

FIG. 5-21. Sources and circuits for which the same current flows into the black box.

if we interpret multiplication by s as the derivative[1] and $1/s$ as the definite integral.

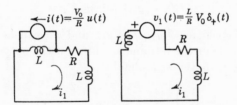

FIG. 5-22. Equivalent circuits for Fig. 5-6b.

Example 3. Let us find the current in Fig. 5-6b using the equivalence shown in Fig. 5-22. According to (5-27)

$$\mathcal{L}[v_1(t)] = LsI(s) = \frac{L}{R} V_0$$
$$v_1(t) = \frac{L}{R} V_0 \delta_+(t)$$

(5-29)

The transformed loop equation is

$$(2Ls + R)I_1 = - \frac{L}{R} V_0$$
$$I_1 = - \frac{L}{R} V_0 \frac{1}{2Ls + R} = - \frac{V_0}{2R} \frac{1}{s + (R/2L)}$$

(5-30)

which checks with (5-5). In this example the exchange of the sources resulted in some saving of work.

----

[1] Starting with

$$\mathcal{L}\left[\frac{df}{dt}\right] = sF(s) - f(0)$$

we have

$$\mathcal{L}^{-1}[sF(s)] = \frac{df}{dt} + f(0)\delta_+(t)$$

5-4. Four-terminal Networks

In Sec. 5-2 we discussed network properties as seen at a pair of terminals. Let us now extend the idea of describing a network by external measurements to the case where two pairs of terminals are brought out. We now suppose that we make measurements on the black box shown in Fig. 5-23.[1]

FIG. 5-23. Four-terminal network.

Because of the superposition property of a linear network (Sec. 1-3), we can write the currents as sums of terms involving the applied voltages (it is assumed that there are no sources inside the black box). The transformed equations are

$$I_1 = y_{11}V_1 + y_{12}V_2$$
$$I_2 = y_{21}V_1 + y_{22}V_2$$

$$(5\text{-}31)$$

The admittance coefficients y_{11}, y_{22}, y_{12}, and y_{21} can be determined by measurements made at the terminals. First, if the output terminals are shorted, then V_2 in (5-31) is zero and

$$y_{11} = \frac{I_1}{V_1} \Bigg\}$$
$$y_{21} = \frac{I_2}{V_1} \Bigg\}$$
output terminals shorted $(5\text{-}32)$

Next the input terminals are shorted, making $v_1 = 0$, and we have

$$y_{22} = \frac{I_2}{V_2} \Bigg\}$$
$$y_{12} = \frac{I_1}{V_2} \Bigg\}$$
input terminals shorted $(5\text{-}33)$

Because of the method of measurement the ys are called short-circuit parameters. It can be shown[2] that $y_{12} = y_{21}$, so that three ys are sufficient to characterize a four-terminal network.

In the same way as the ys were developed, a set of impedances can be found to describe the black box. We write instead of (5-31)

$$V_1 = z_{11}I_1 + z_{12}I_2$$
$$V_2 = z_{21}I_1 + z_{22}I_2$$

$$(5\text{-}34)$$

[1] Notice the direction of the current i_2. It is customary to define network parameters for this current direction which differs from the clockwise convention we have been using.

[2] E. A. Guillemin, "Communication Networks," vol. 2, chap. 4, John Wiley & Sons, Inc., New York, 1935.

The zs are called open-circuit parameters because they can be measured by first opening the output terminals so that $i_2 = 0$ and

$$z_{11} = \frac{V_1}{I_1} \Bigg\} \quad \text{output terminals open} \qquad (5\text{-}35)$$
$$z_{21} = \frac{V_2}{I_1} \Bigg\}$$

Then the input terminals are opened, making $i_1 = 0$, so that

$$z_{22} = \frac{V_2}{I_2} \Bigg\} \quad \text{input terminals open} \qquad (5\text{-}36)$$
$$z_{12} = \frac{V_1}{I_2} \Bigg\}$$

Again, $z_{12} = z_{21}$, so that three parameters are sufficient. It is worthwhile to note that the zs and ys are not reciprocals of one another. There are several other sets of parameters which can be used to describe a four-terminal network, but we shall not discuss them here.[1]

Example 1: T Networks. Let us find the impedance parameters for the network shown in Fig. 5-24. From (5-35) and (5-36) we have immediately

$$z_{11} = Z_1 + Z_2$$
$$z_{22} = Z_3 + Z_2 \qquad (5\text{-}37)$$

Now either z_{12} or z_{21} can be found, since the two are equal. With no

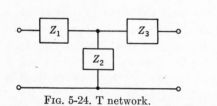

FIG. 5-24. T network.

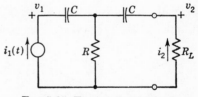

FIG. 5-25. T network with load.

current flowing at the output terminals there will be no voltage drop across Z_3. Then from (5-35)

$$z_{21} = \frac{\mathcal{L}\,[\text{voltage across } Z_2]}{\mathcal{L}\,[\text{input current}]}$$
$$= \frac{Z_2 I_1}{I_1} = Z_2 \qquad (5\text{-}38)$$

Suppose we wish to find the voltage across R_L in Fig. 5-25 due to the application of current $i_1(t)$ at the input terminals. From (5-37) and

[1] *Ibid.*

(5-38) we obtain the zs from the T network:

$$z_{11} = \frac{1}{Cs} + R$$

$$z_{22} = \frac{1}{Cs} + R \tag{5-39}$$

$$z_{12} = R$$

These are substituted in (5-34):

$$V_1 = \frac{RCs + 1}{Cs} I_1 + RI_2 \tag{5-40}$$

$$V_2 = RI_1 + \frac{RCs + 1}{Cs} I_2 \tag{5-41}$$

The load resistor R_L across the output terminals forces a relation between I_2 and V_2:

$$I_2 = -\frac{V_2}{R_L} \tag{5-42}$$

The voltage V_2 can now be obtained from (5-41) by substituting (5-42):

$$V_2 = RI_1 - \frac{RCs + 1}{Cs} \frac{V_2}{R_L}$$

$$= \frac{RR_LCs}{(R_L + R)Cs + 1} I_1 \tag{5-43}$$

We could have used loop analysis with about the same amount of labor to find (5-43). However, when changes in the load or input impedance are of interest, this method of analysis has an advantage, since it treats part of the network (in this case the T) as a unit.

Example 2: Lattice Network. A network of special interest in network theory is the *lattice* network shown in Fig. 5-26. The symmetrical lattice is the most general symmetrical four-terminal network. A symmetrical network has, by definition, the property that $z_{11} = z_{22}$, and any such network can be replaced by an equivalent symmetrical lattice.[1]

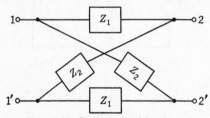

FIG. 5-26. Symmetrical lattice.

To analyze the lattice of Fig. 5-26, it is convenient to rearrange the network as shown in Fig. 5-27.

[1] *Ibid.*, chap. 10.

We can find $z_{22} = z_{11}$ by measuring the impedance seen from the terminals 2-2′ with 1-1′ open.   It is not hard to show that this is

$$z_{11} = \tfrac{1}{2}(Z_1 + Z_2) \tag{5-44}$$

The transfer impedance can be obtained by applying a current i_1 at terminals 1-1′ and determining the voltage at 2-2′ as shown in Fig. 5-28.

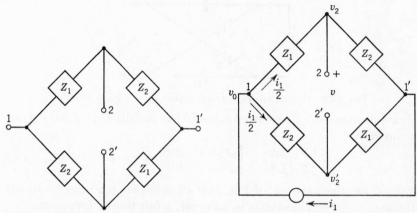

FIG. 5-27. Bridge network obtained by rearranging the network of Fig. 5-26.

FIG. 5-28. Determination of z_{12} for lattice.

Since the impedances in the parallel branches are equal, the current will split in equal parts as shown.   The transformed voltage at 2 will be

$$V_2(s) = V_0(s) - Z_1(s) \frac{I_1(s)}{2} \tag{5-45}$$

and at 2′

$$V_{2'}(s) = V_0(s) - Z_2(s) \frac{I_1(s)}{2} \tag{5-46}$$

Then $$V(s) = V_2(s) - V_{2'}(s) = \tfrac{1}{2}[Z_2(s) - Z_1(s)]I_1(s) \tag{5-47}$$

so that $$z_{12} = \frac{V(s)}{I_1(s)} = \frac{1}{2}(Z_2 - Z_1) \tag{5-48}$$

An interesting lattice network is shown in Fig. 5-29.   Now since both expressions of (5-35) hold for the open-circuited output case shown in the figure, we can write

$$\frac{V_2}{V_1} = \frac{V_2/I_1}{V_1/I_1} = \frac{z_{12}}{z_{11}} \tag{5-49}$$

so that from (5-44) and (5-48)

$$\frac{V_2(s)}{V_1(s)} = \frac{z_{12}(s)}{z_{11}(s)} = \frac{Z_2 - Z_1}{Z_2 + Z_1} \tag{5-50}$$

In the example

$$\frac{V_2(s)}{V_1(s)} = \frac{R - (1/Cs)}{R + (1/Cs)} = \frac{s - (1/RC)}{s + (1/RC)} \tag{5-51}$$

It will be shown in Chap. 9 that the ratio $V_2(s)/V_1(s)$ (called a *transfer function*) can be related to the steady-state response of the network when

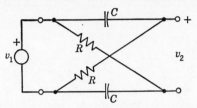

Fig. 5-29. All-pass lattice network driven by a voltage source.

v_1 is a unit sinusoid.   The output sinusoid has amplitude $|V_2(j\omega)/V_1(j\omega)|$. In this case

$$\left| \frac{V_2(j\omega)}{V_1(j\omega)} \right| = \left[\frac{\omega^2 + (1/R^2C^2)}{\omega^2 + (1/R^2C^2)} \right]^{\frac{1}{2}} = 1 \tag{5-52}$$

That is, for the network of Fig. 5-29 all sinusoids are passed without. attenuation.   The phase shift is, however, a function of frequency.

The ratio $V_2(s)/V_1(s)$ has a zero in the right half plane at $s = 1/RC$ Functions like (5-51) with zeros in the right half plane are called *non-minimum phase* functions.[1]

PROBLEMS

5-1. Find the transform of the indicated current for each network.   The switches are closed at $t = 0$.   The networks are assumed to be in equilibrium prior to switching. The sources produce constant voltages and currents.

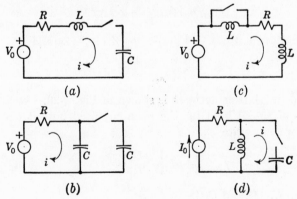

PROB. 5-1

[1] See H. W. Bode, "Network Analysis and Feedback Amplifier Design," D. Van Nostrand Company, Inc., Princeton, N.J., 1945.

5-2. Same as Prob. 5-1, except that switches in networks below are opened at $t = 0$.

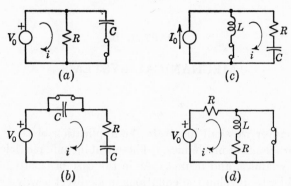

PROB. 5-2

5-3. Show that circuits of Figs. 5-16 and 5-17 are equivalent as far as load voltage is concerned.

5-4. Show by exchange of sources that the current in R_L is the same for both networks.

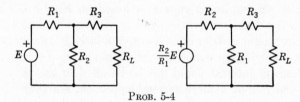

PROB. 5-4

5-5. Verify that the circuits in Figs. 5-19, 5-20, and 5-21 are really equivalent to Fig. 5-18 as far as the black box is concerned.

5-6. Consider a four-terminal network consisting of a general three-loop network with one source in each of two loops. Show that $y_{12} = y_{21}$ where the ys are defined by (5-31).

5-7. Find z_{11} and z_{12} for the networks shown.

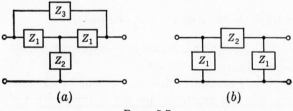

PROB. 5-7

CHAPTER 6

MECHANICAL SYSTEMS

In this chapter we shall illustrate the application of the $\mathcal{L}$-transform method to problems of mechanics. These will include particle dynamics, rigid-body dynamics, and beams. As in the case of electrical networks, our aim will be to develop the equations in as direct a way as possible.

6-1. Dynamics—Lagrange's Equations

In studying electrical networks we started with three kinds of elements —resistors, capacitors, and inductors—and connected these together in an arbitrary way. The currents flowing in the resulting network are then found by solving linear equations based on Kirchhoff's laws. We could follow the same pattern in dealing with mechanical systems, but in so doing we would restrict ourselves to a relatively small class of problems—those involving the connection of certain kinds of elements. Many other problems of interest could not be handled in this way. For this reason we shall adopt a more general approach, starting with the law rather than a specification of elements. We shall find that nonlinear equations are the rule and that we shall usually have to make approximations before attempting to solve by $\mathcal{L}$ transforms.

We shall now *postulate* that the dynamic behavior of any mechanical system is described by a set of equations of the form

$$\frac{d}{dt}\left(\frac{\partial T}{\partial \dot{q}_i}\right) - \frac{\partial T}{\partial q_i} = Q_i \tag{6-1}$$

These are called *Lagrange's equations.*[1] In (6-1),

T = kinetic energy

q_i = one of the coordinates describing position of system

Q_i = generalized force

More explanation is needed. First, the qs can be any set of quantities (like displacements and angles) which define the position of the system. They are called *generalized coordinates.*

---

[1] Lagrange's equations can be derived from Newton's laws. See, for example, E. T. Whittaker, "Analytical Dynamics," 4th ed., pp. 34ff., Dover Publications, New York, 1944.

The work done in making a small change in the coordinates is written

$$\delta W = Q_1\,\delta q_1 + \cdots + Q_i\,\delta q_i + \cdots$$

This expression defines the generalized forces Q_i.

The great advantage of using Lagrange's equations lies in the fact that (6-1) is true no matter how the coordinates are chosen. It is true, for example, in a rotating coordinate system or a fixed one. In contrast, while we might write Newton's law

$$Force = mass \times acceleration \tag{6-2}$$

in an inertial reference system, we must write it

$$Force = (mass \times acceleration) - (Coriolis\ force)$$
$$- (centrifugal\ force) \tag{6-3}$$

in a rotating set of coordinates.

We shall illustrate the application of Lagrange's equations by examples.

Example 1: Linear Translational System. Let us consider the system shown in Fig. 6-1. We shall take as our coordinate x, the distance of the mass from the rest position. The kinetic energy is

$$T = \tfrac{1}{2}M\dot{x}^2 \tag{6-4}$$

If the spring provides a force proportional to its extension or compression, then that force can be written

$$F_K = -Kx \tag{6-5}$$

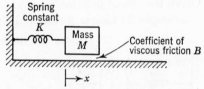

Fig. 6-1. Simple translational system.

Viscous friction is defined as providing a force proportional to velocity:

$$F_B = -B\dot{x} \tag{6-6}$$

The work done by a small change in the x coordinate from an arbitrary position is the product of force and distance δx:

$$\delta W = -Kx\,\delta x - B\dot{x}\,\delta x \tag{6-7}$$

We now identify the generalized force [in this case identical with the sum of forces defined in (6-5) and (6-6)]:

$$Q = -Kx - B\dot{x} \tag{6-8}$$

Lagrange's equation for the system is

$$\frac{d}{dt}\left(\frac{\partial T}{\partial \dot{x}}\right) - \frac{\partial T}{\partial x} = -Kx - B\dot{x} \tag{6-9}$$

From (6-4)

$$\frac{d}{dt}\left(\frac{\partial T}{\partial \dot{x}}\right) = \frac{d}{dt} M\dot{x} = M\ddot{x}$$

$$\frac{\partial T}{\partial x} = 0 \tag{6-10}$$

Notice that $\dot{x}$ and x are treated as independent variables in taking the partial derivatives. Equation (6-9) is now written

$$M\ddot{x} = -Kx - B\dot{x} \tag{6-11}$$

which is the same result that we would have obtained by applying Newton's laws. This example serves as a verification *in this case* of our postulate that Lagrange's equations describe the behavior of mechanical systems. The utility of the Lagrange method will become apparent in more complicated examples.

Transforming (6-11) with $\mathcal{L}[x(t)] \triangleq X(s)$, we have

$$Ms^2X(s) - Msx(0) - M\dot{x}(0) = -KX(s) - BsX(s) + Bx(0)$$

$$X(s) = \frac{Msx(0) + M\dot{x}(0) + Bx(0)}{Ms^2 + Bs + K} \tag{6-12}$$

Given the initial position and velocity, $x(t)$ can be obtained by inversion.

Example 2: Linear Rotational System. Now let us find the equation

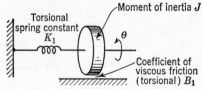

FIG. 6-2. Rotational system.

of motion for the system shown in Fig. 6-2. In this case we choose θ, the angle measured from the rest position, as the coordinate. The kinetic energy is

$$T = \tfrac{1}{2}J\dot{\theta}^2 \tag{6-13}$$

The spring is assumed to provide a torque L proportional to θ:

$$L_K = -K_1\theta \tag{6-14}$$

The viscous friction provides a torque proportional to $\dot{\theta}$:

$$L_B = -B_1\dot{\theta} \tag{6-15}$$

The work done in rotating $\delta\theta$ from an arbitrary position is

$$\delta W = -K_1\theta\,\delta\theta - B_1\dot{\theta}\,\delta\theta \tag{6-16}$$

so

$$Q = -K_1\theta - B_1\dot{\theta} \tag{6-17}$$

and Lagrange's equation for the system is

$$\frac{d}{dt}\left(\frac{\partial T}{\partial \dot{\theta}}\right) - \frac{\partial T}{\partial \theta} = -K_1\theta - B_1\dot{\theta} \tag{6-18}$$

From (6-13)

$$\frac{d}{dt}\left(\frac{\partial T}{\partial \dot\theta}\right) = \frac{d}{dt}J\dot\theta = J\ddot\theta$$

$$\frac{\partial T}{\partial \theta} = 0 \tag{6-19}$$

So finally we write

$$J\ddot\theta = -K_1\theta - B_1\dot\theta \tag{6-20}$$

This equation is of the same form as (6-11) and can be transformed in the same way.

Example 3: Coupled Translational System. The two masses in Fig. 6-3 are connected to each other by a spring; the mass M_2 is restrained by viscous friction as shown. The two coordinates needed to describe the position of this system are taken as x_1 and x_2, the displacements of the masses from their rest positions. The kinetic energy is

$$T = \tfrac{1}{2}(M_1\dot x_1{}^2 + M_2\dot x_2{}^2) \tag{6-21}$$

The work done in displacing M_1, while holding x_2 constant, is

$$\delta W_1 = -K_1x_1\,\delta x_1 - K_2(x_1 - x_2)\,\delta x_1 \tag{6-22}$$

The work done in moving M_2 is

$$\delta W_2 = -K_2(x_2 - x_1)\,\delta x_2 - B\dot x_2\,\delta x_2 \tag{6-23}$$

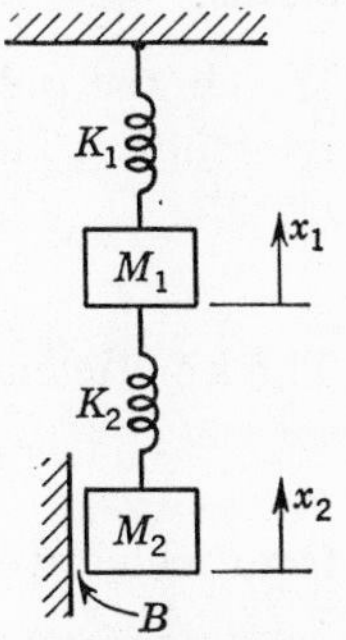

Fig. 6-3. Coupled system.

The force of gravity is just canceled by the force provided by the springs in holding the masses in their equilibrium positions.

There are two equations:

$$\frac{d}{dt}\left(\frac{\partial T}{\partial \dot x_1}\right) - \frac{\partial T}{\partial x_1} = -K_1x_1 - K_2(x_1 - x_2)$$

$$\frac{d}{dt}\left(\frac{\partial T}{\partial \dot x_2}\right) - \frac{\partial T}{\partial x_2} = -K_2(x_2 - x_1) - B\dot x_2 \tag{6-24}$$

Using (6-21) these become

$$M_1\ddot x_1 + (K_1 + K_2)x_1 - K_2x_2 = 0$$
$$-K_2x_1 + M_2\ddot x_2 + B\dot x_2 + K_2x_2 = 0 \tag{6-25}$$

Transforming with $X_1(s) \triangleq \mathcal{L}[x_1(t)]$, etc., we have

$$\begin{bmatrix} M_1s^2 + (K_1 + K_2) & -K_2 \\ -K_2 & M_2s^2 + Bs + K_2 \end{bmatrix}\begin{bmatrix} X_1(s) \\ X_2(s) \end{bmatrix}$$
$$= \begin{bmatrix} M_1sx_1(0) + M_1\dot x_1(0) \\ M_2sx_2(0) + M_2\dot x_2(0) + Bx(0) \end{bmatrix} \tag{6-26}$$

Example 4: Particle in a Rotating Coordinate System. So far the examples have led to linear equations which could be transformed. This

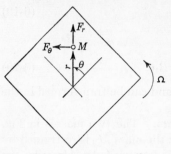

is partly the result of the choice of coordinates and partly because the elements—springs and frictional devices—were assumed to have certain properties. To show how the Lagrange method simplifies writing the equations in complicated problems, we shall find the equations of motion for a particle moving in a rotating coordinate system. The equation resulting will be nonlinear.

FIG. 6-4. Rotating coordinate system.

The coordinate system in Fig. 6-4 rotates with constant angular velocity Ω. We choose polar coordinates r and θ measured in the rotating frame of reference. The velocity of the mass M has two components:

$$V_r = \dot{r}$$
$$V_\theta = r(\dot{\theta} + \Omega) \tag{6-27}$$

The kinetic energy is

$$T = \tfrac{1}{2}M[\dot{r}^2 + r^2(\dot{\theta}^2 + 2\dot{\theta}\Omega + \Omega^2)] \tag{6-28}$$

from which we write

$$\frac{d}{dt}\left(\frac{\partial T}{\partial \dot{r}}\right) = \frac{d}{dt}M\dot{r} = M\ddot{r}$$
$$\frac{\partial T}{\partial r} = Mr(\dot{\theta}^2 + 2\dot{\theta}\Omega + \Omega^2) \tag{6-29}$$

$$\frac{d}{dt}\left(\frac{\partial T}{\partial \dot{\theta}}\right) = \frac{d}{dt}[Mr^2\dot{\theta} + Mr^2\Omega] = 2Mr\dot{r}\dot{\theta} + Mr^2\ddot{\theta} + 2Mr\dot{r}\Omega$$
$$\frac{\partial T}{\partial \theta} = 0 \tag{6-30}$$

We assume that forces F_r and F_θ are applied to M. The work done by these forces in a small displacement is

$$\delta W = F_r\,\delta r + F_\theta\,r\delta\theta \tag{6-31}$$

The Lagrange equations are written from (6-29), (6-30), and (6-31):

$$M\ddot{r} - Mr\dot{\theta}^2 - 2Mr\dot{\theta}\Omega - Mr\Omega^2 = F_r$$
$$Mr\ddot{\theta} + 2M\dot{r}\dot{\theta} + 2M\dot{r}\Omega = F_\theta \tag{6-32}$$

The equations can be rewritten as

$$M\ddot{r} - Mr\dot{\theta}^2 = F_r + 2Mr\dot{\theta}\Omega + Mr\Omega^2$$
$$Mr\ddot{\theta} + 2M\dot{r}\dot{\theta} = F_\theta - 2M\dot{r}\Omega \tag{6-33}$$

The terms $2Mr\dot{\theta}\Omega$ and $-2M\dot{r}\Omega$ are components of the Coriolis force. The term $Mr\Omega^2$ is the centrifugal force. These equations are nonlinear and cannot be transformed in the usual way. In the next sections we shall discuss methods of linearizing equations like these.

6-2. Perturbation Equations—The Motion of a Satellite

To illustrate the perturbation method for linearization of nonlinear equations, we shall study the following problem.

A satellite is following a circular orbit about the earth. We ask what the effect is of a small displacement away from the orbit.

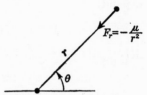

The technique is first to write the equations for the undisturbed motion and then assume a solution consisting of the steady solution plus a perturbation quantity—in this case the

Fig. 6-5. Particle under gravitational force.

deviation from the circular orbit. The resulting *perturbation equation* will be linear, and we can apply the $\mathcal{L}$ transform to it.

Suppose a particle of mass M is attracted to another particle by a gravitational force

$$F_r = -\frac{\mu}{r^2} \tag{6-34}$$

as shown in Fig. 6-5. The kinetic energy is

$$T = \tfrac{1}{2}M(\dot{r}^2 + r^2\dot{\theta}^2) \tag{6-35}$$

and Lagrange's equations can be taken from (6-33) with $\Omega = 0$:

$$M\ddot{r} - Mr\dot{\theta}^2 = -\frac{\mu}{r^2}$$
$$Mr\ddot{\theta} + 2M\dot{r}\dot{\theta} = 0 \tag{6-36}$$

We observe that the second equation, when written

$$\frac{1}{r}\frac{d}{dt}(Mr^2\dot{\theta}) = 0 \tag{6-37}$$

is a statement that the angular momentum h is constant

$$h = Mr^2\dot{\theta} = \text{constant} \tag{6-38}$$

A circular orbit in which $r = r_0$, a constant, is one possible solution of (6-36). Then the angular velocity $\dot{\theta}_0$ will be given by the solution of

$$Mr_0\dot{\theta}_0{}^2 = \frac{\mu}{r_0{}^2} \tag{6-39}$$

Let us now determine the effect of a small displacement of the particle from the circular orbit. We write

$$r = r_0 + r_1 \qquad \dot{r} = \dot{r}_1$$
$$\theta = \theta_0 + \theta_1 \qquad \ddot{\theta} = \ddot{\theta}_1 \qquad (6\text{-}40)$$

where r_1 and $\dot{\theta}_1$ are small quantities and $\dot{\theta}_0$ and r_0 are the constants for a circular orbit. We now substitute in the first equation of (6-36) and expand:

$$M\ddot{r}_1 - M(r_0 + r_1)(\dot{\theta}_0 + \dot{\theta}_1)^2 = -\frac{\mu}{(r_0 + r_1)^2}$$

$$M\ddot{r}_1 - Mr_0\dot{\theta}_0{}^2 - Mr_0\dot{\theta}_1{}^2 - Mr_1\dot{\theta}_0{}^2 - Mr_1\dot{\theta}_1{}^2$$
$$-2Mr_0\dot{\theta}_0\dot{\theta}_1 - 2Mr_1\dot{\theta}_0\dot{\theta}_1 = -\frac{\mu}{r_0{}^2[1 + (r_1/r_0)]^2} \qquad (6\text{-}41)$$

Now since r_1 and $\dot{\theta}_1$ are assumed small, we neglect any terms involving squares or products of these quantities. Also r_1/r_0 is a small number, and the term on the right can be written

$$-\frac{\mu}{r_0{}^2[1 + (r_1/r_0)]^2} \cong -\frac{\mu}{r_0{}^2}\left[1 - 2\frac{r_1}{r_0} + 3\left(\frac{r_1}{r_0}\right)^2 - \cdots\right]$$
$$\cong \frac{-\mu[1 - 2(r_1/r_0)]}{r_0{}^2} = -\frac{\mu}{r_0{}^2} + 2\frac{r_1}{r_0{}^3}\mu \qquad (6\text{-}42)$$

We now write (6-41)

$$M\ddot{r}_1 - 2Mr_0\dot{\theta}_1\dot{\theta}_0 - Mr_1\dot{\theta}_0{}^2 - \left[Mr_0\dot{\theta}_0{}^2 - \frac{\mu}{r_0{}^2}\right] = 2\mu\frac{r_1}{r_0{}^3} \qquad (6\text{-}43)$$

The term in the brackets is zero by (6-39). The second equation of (6-36) is now written

$$M(r_0 + r_1)\ddot{\theta}_1 + 2M\dot{r}_1(\dot{\theta}_0 + \dot{\theta}_1) = 0 \qquad (6\text{-}44)$$

which becomes, when we neglect products of $\dot{\theta}_1$ and r_1 and their derivatives (which are also assumed small),

$$Mr_0\ddot{\theta}_1 + 2M\dot{r}_1\dot{\theta}_0 = 0 \qquad (6\text{-}45)$$

The linear equations for small motions superimposed on a steady circular orbit are from (6-43) and (6-45)

$$M\ddot{r}_1 - \left(M\dot{\theta}_0{}^2 + \frac{2\mu}{r_0{}^3}\right)r_1 - 2Mr_0\dot{\theta}_0\dot{\theta}_1 = 0$$
$$2M\dot{\theta}_0\dot{r}_1 + Mr_0\ddot{\theta}_1 = 0 \qquad (6\text{-}46)$$

The coefficient of r_1 in the first equation is $-3M\dot{\theta}_0{}^2$ by (6-39). Transforming, with $\bar{r}_1(s) \triangleq \mathcal{L}[r_1(t)]$, etc., we have

$$
\begin{bmatrix} Ms^2 - 3M\dot{\theta}_0{}^2 & -2Mr_0\dot{\theta}_0s \\ 2M\dot{\theta}_0s & Mr_0s^2 \end{bmatrix} \begin{bmatrix} \bar{r}_1(s) \\ \bar{\theta}_1(s) \end{bmatrix}
$$
$$
= \begin{bmatrix} Msr_1(0) + M\dot{r}_1(0) - 2Mr_0\dot{\theta}_0\theta_1(0) \\ 2M\dot{\theta}_0r_1(0) + Mr_0s\theta_1(0) + Mr_0\dot{\theta}_1(0) \end{bmatrix} \quad (6\text{-}47)
$$

Taking the disturbance to be an outward displacement $r_1(0)$, with all other initial conditions zero, we have for $\bar{r}_1(s)$

$$
\bar{r}_1(s) = \frac{\begin{vmatrix} Msr_1(0) & -2Mr_0\dot{\theta}_0s \\ 2M\dot{\theta}_0r_1(0) & Mr_0s^2 \end{vmatrix}}{\begin{vmatrix} Ms^2 - 3M\dot{\theta}_0{}^2 & -2Mr_0\dot{\theta}_0s \\ 2M\dot{\theta}_0s & Mr_0s^2 \end{vmatrix}} = \frac{r_0M^2sr_1(0)(s^2 + 4\,\dot{\theta}_0{}^2)}{r_0M^2s^2(s^2 + \dot{\theta}_0{}^2)}
$$

$$
= r_1(0)\frac{s^2 + 4\dot{\theta}_0{}^2}{s(s^2 + \dot{\theta}_0{}^2)}
$$

$$
= r_1(0)\left[\frac{4}{s} - \frac{3s}{s^2 + \dot{\theta}_0{}^2}\right] \quad (6\text{-}48)
$$

The inverse transform is

$$
r_1(t) = r_1(0)[4 - 3\cos\dot{\theta}_0 t]
$$

In Fig. 6-6 are shown plots of r_1 and of the new orbit.

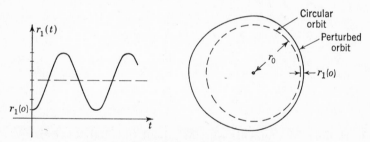

FIG. 6-6. Perturbed motion of a satellite.

6-3. Gyroscopes

In this section we shall study some of the simpler gyroscope configurations. Figure 6-7 shows a set of gimbals with a symmetrical body mounted at the center. The angles ϕ, θ, and ψ will be used as coordinates, and the components of body angular velocity are as shown. The component ω_2 lies along the θ-gimbal axis, and component ω_1 is at right angles to it and to ω_3.

The body has equal moments of inertia J_1 about the ω_1 and ω_2 axes and moment of inertia J_2 about the axis of symmetry. In order to find the kinetic energy, we shall need to know the angular velocities about these axes in terms of the coordinate angles.[1] The angular velocities are written by inspection:

$$\omega_1 = -\dot{\phi}\sin\theta$$
$$\omega_2 = \dot{\theta} \tag{6-49}$$
$$\omega_3 = \dot{\psi} + \dot{\phi}\cos\theta$$

The kinetic energy (assuming the gimbals are weightless) is

$$T = \tfrac{1}{2}J_1(\dot{\phi}^2\sin^2\theta + \dot{\theta}^2) + \tfrac{1}{2}J_2(\dot{\psi} + \dot{\phi}\cos\theta)^2 \tag{6-50}$$

We now take the derivatives necessary for Lagrange's equations.

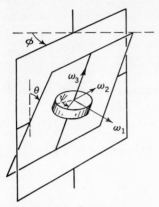

Fig. 6-7. Gyroscope.

$$\frac{d}{dt}\left(\frac{\partial T}{\partial\dot{\phi}}\right) = \frac{d}{dt}[J_1\dot{\phi}\sin^2\theta + J_2(\dot{\psi} + \dot{\phi}\cos\theta)\cos\theta]$$
$$= J_1\ddot{\phi}\sin^2\theta + 2J_1\dot{\phi}\dot{\theta}\sin\theta\cos\theta + J_2\ddot{\psi}\cos\theta + J_2\ddot{\phi}\cos^2\theta$$
$$- J_2\dot{\phi}\dot{\theta}\sin\theta\cos\theta - J_2\dot{\theta}(\dot{\psi} + \dot{\phi}\cos\theta)\sin\theta$$

$$\frac{\partial T}{\partial\phi} = 0 \tag{6-51}$$

$$\frac{d}{dt}\left(\frac{\partial T}{\partial\dot{\theta}}\right) = \frac{d}{dt}J_1\dot{\theta} = J_1\ddot{\theta}$$
$$\frac{\partial T}{\partial\theta} = J_1\dot{\phi}^2\sin\theta\cos\theta - J_2(\dot{\psi} + \dot{\phi}\cos\theta)\dot{\phi}\sin\theta \tag{6-52}$$

$$\frac{d}{dt}\left(\frac{\partial T}{\partial\dot{\psi}}\right) = \frac{d}{dt}J_2(\dot{\psi} + \dot{\phi}\cos\theta) = J_2\ddot{\psi} + J_2\ddot{\phi}\cos\theta - J_2\dot{\phi}\dot{\theta}\sin\theta$$
$$\frac{\partial T}{\partial\psi} = 0 \tag{6-53}$$

The work done by moments supplied about the gimbal axes is

$$\delta W = M_\phi\,\delta\phi + M_\theta\,\delta\theta + M_\psi\,\delta\psi \tag{6-54}$$

The Lagrangian equations can now be written

$$J_1\ddot{\phi}\sin^2\theta + J_2\ddot{\psi}\cos\theta + J_2\ddot{\phi}\cos^2\theta + 2(J_1 - J_2)\dot{\phi}\dot{\theta}\sin\theta\cos\theta$$
$$- J_2\dot{\theta}\dot{\psi}\sin\theta = M_\phi$$
$$J_1\ddot{\theta} - (J_1 - J_2)\dot{\phi}^2\sin\theta\cos\theta + J_2\dot{\psi}\dot{\phi}\sin\theta = M_\theta \tag{6-55}$$
$$J_2\ddot{\psi} + J_2\ddot{\phi}\cos\theta - J_2\dot{\phi}\dot{\theta}\sin\theta = M_\psi$$

[1] The angles ϕ, θ, and ψ are called Euler angles. See, for example, J. L. Synge and B. A. Griffith, "Principles of Mechanics," 2d ed., p. 288, McGraw-Hill Book Company, New York, 1949, or Whittaker, *op. cit.*, p. 9.

We shall now make some assumptions in order to simplify these nonlinear equations. First, let us assume that $M_\psi = 0$ and that the body is rotating at a constant speed $\dot\psi_0$. Furthermore, we assume that there is no friction in the θ gimbal, so that $M_\theta = 0$. The equations now are

$$
\begin{aligned}
J_1\ddot\phi \sin^2\theta + J_2\ddot\phi \cos^2\theta + 2(J_1 - J_2)\dot\phi\dot\theta \sin\theta\cos\theta \\
- J_2\dot\theta\dot\psi_0 \sin\theta = M_\phi \\
J_1\ddot\theta - (J_1 - J_2)\dot\phi^2 \sin\theta\cos\theta + J_2\dot\psi_0\dot\phi \sin\theta = 0 \\
J_2\ddot\phi \cos\theta - J_2\dot\phi\dot\theta \sin\theta = 0
\end{aligned}
\tag{6-56}
$$

Let us now orient the gimbals so that $\theta \cong \pi/2$. Then for small deviations θ_1 from this position we have

$$
\begin{aligned}
\theta = \frac{\pi}{2} + \theta_1 \qquad \dot\theta = \dot\theta_1 \\
\sin\theta \cong 1 \\
\cos\theta \cong -\theta_1
\end{aligned}
\tag{6-57}
$$

Equations (6-56) now become

$$
\begin{aligned}
J_1\ddot\phi + J_2\ddot\phi\theta_1{}^2 + 2(J_1 - J_2)\dot\phi\dot\theta_1(-\theta_1) - J_2\dot\theta_1\dot\psi_0 = M_\phi \\
J_1\ddot\theta_1 - (J_1 - J_2)\dot\phi^2(-\theta_1) + J_2\dot\psi_0\dot\phi = 0 \\
J_2\ddot\phi(-\theta_1) - J_2\dot\phi\dot\theta_1 = 0
\end{aligned}
\tag{6-58}
$$

We now drop all terms containing products of the small quantities $\ddot\phi$, $\dot\phi$, $\dot\theta_1$, and θ_1. The third equation vanishes identically, and the first two become

$$
\begin{aligned}
J_1\ddot\phi - J_2\dot\psi_0\dot\theta_1 = M_\phi \\
J_2\dot\psi_0\dot\phi + J_1\ddot\theta_1 = 0
\end{aligned}
\tag{6-59}
$$

These linear equations describe the low-amplitude motion when a moment M_ϕ is applied and θ is near 90°. The constant term

$$
J_2\dot\psi_0 \triangleq h
\tag{6-60}
$$

is the angular momentum of the body. Writing $\mathcal{L}[\theta_1(t)] \triangleq \bar\theta_1(s)$, etc., we write the transformed equations, with initial conditions zero,

$$
\begin{bmatrix} J_1s^2 & -hs \\ hs & J_1s^2 \end{bmatrix}
\begin{bmatrix} \bar\phi(s) \\ \bar\theta_1(s) \end{bmatrix}
=
\begin{bmatrix} \bar M_\phi(s) \\ 0 \end{bmatrix}
\tag{6-61}
$$

The transformed angle $\bar\theta_1(s)$ is

$$
\begin{aligned}
\bar\theta_1(s) &= \frac{\begin{vmatrix} J_1s^2 & \bar M_\phi(s) \\ hs & 0 \end{vmatrix}}{\begin{vmatrix} J_1s^2 & -hs \\ hs & J_1s^2 \end{vmatrix}} = \frac{-\bar M_\phi(s)hs}{s^2(J_1{}^2s^2 + h^2)} \\
&= \frac{-\bar M_\phi(s)(h/J_1{}^2)}{s[s^2 + (h^2/J_1{}^2)]} = -\bar M_\phi(s)\frac{1}{h}\left[\frac{1}{s} - \frac{s}{s^2 + (h^2/J_1{}^2)}\right]
\end{aligned}
\tag{6-62}
$$

The transform of ϕ is

$$\bar{\phi}(s) = \frac{\bar{M}_\phi(s) J_1 s^2}{s^2 (J_1^2 s^2 + h^2)} = \frac{\bar{M}_\phi(s)(1/J_1)}{s^2 + (h^2/J_1^2)} \qquad (6\text{-}63)$$

If M_ϕ is an impulse function of unit area, then

$$\theta_1(t) = -\frac{1}{h} + \frac{1}{h} \cos\left(\frac{h}{J_1}\right) t$$

$$\phi(t) = \frac{1}{h} \sin\left(\frac{h}{J_1}\right) t \qquad (6\text{-}64)$$

This result shows that if the outer gimbal is given an impulsive moment, the system will oscillate indefinitely at a frequency h/J_1 radians per second and the inner gimbal will be displaced an amount proportional to $1/h$. If the angular momentum h is large, the frequency will be high and the displacement of the inner gimbal small, which is to say that the system will tend to become rigid.

Floated Gyro. In the absence of torques about the gimbal axes the gyro of Fig. 6-7 will maintain a fixed orientation with respect to inertial space. Mounted in an aircraft, the gimbal angles can be used as measures of orientation. The effects of torques caused by friction in gimbal bearings can be reduced by making h large. Another approach is the reduction of friction torques by flotation of the inner gimbal to reduce loading on the bearings.

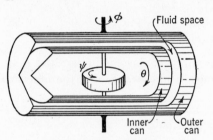

FIG. 6-8. Floated gyro.

Figure 6-8 shows how this can be done. The space between cans is filled with a fluid of sufficient density to float the inner can and thus reduce bearing loads. The fluid has vicosity, so that a torque is produced by the rotation of the inner gimbal. In Eq. (6-55) this torque appears as a moment proportional to $\dot{\theta}$:

$$M_\theta = -c\dot{\theta} \qquad (6\text{-}65)$$

After making the small angle approximations with θ approximately $\pi/2$ as before, we write (6-59) as

$$J_1 \ddot{\phi} - h\dot{\theta}_1 = M_\phi$$

$$h\dot{\phi} + c\dot{\theta}_1 + J_1 \ddot{\theta}_1 = 0 \qquad (6\text{-}66)$$

Transforming as before, we now have

$$\begin{bmatrix} J_1 s^2 & -hs \\ hs & cs + J_1 s^2 \end{bmatrix} \begin{bmatrix} \bar{\phi}(s) \\ \bar{\theta}_1(s) \end{bmatrix} = \begin{bmatrix} \bar{M}_\phi \\ 0 \end{bmatrix} \qquad (6\text{-}67)$$

Now

$$\bar{\theta}_1(s) = \frac{\begin{vmatrix} J_1 s^2 & \bar{M}_\phi \\ hs & 0 \end{vmatrix}}{\begin{vmatrix} J_1 s^2 & -hs \\ hs & cs + J_1 s^2 \end{vmatrix}} = \frac{-\bar{M}_\phi hs}{s^2(J_1^2 s^2 + J_1 cs + h^2)} \qquad (6\text{-}68)$$

$$\bar{\phi}(s) = \frac{\begin{vmatrix} \bar{M}_\phi & -hs \\ 0 & cs + J_1 s^2 \end{vmatrix}}{s^2(J_1^2 s^2 + J_1 cs + h^2)} = \frac{\bar{M}_\phi(cs + J_1 s^2)}{s^2(J_1^2 s^2 + J_1 cs + h^2)}$$

The ratio of $\bar{\theta}_1(s)$ to $\bar{\phi}(s)$ is

$$\frac{\bar{\theta}_1(s)}{\bar{\phi}(s)} = \frac{-\bar{M}_\phi hs}{\bar{M}_\phi(cs + J_1 s^2)} = \frac{-h/J_1}{s + (c/J_1)}$$

$$\bar{\theta}_1(s) = \frac{-h/J_1}{s + (c/J_1)}\, \bar{\phi}(s) \qquad (6\text{-}69)$$

If $\phi(t)$ is a unit step function, then

$$\bar{\theta}_1(s) = -\frac{h}{J_1}\left\{\frac{1}{s[s + (c/J_1)]}\right\} = -\frac{h}{J_1}\left[\frac{J_1/c}{s} - \frac{J_1/c}{s + (c/J_1)}\right] \qquad (6\text{-}70)$$

and the inverse transform is

$$\theta_1(t) = -\frac{h}{c}[1 - e^{-(c/J_1)t}] \qquad (6\text{-}71)$$

which is plotted in Fig. 6-9. In practice h and c can be made of the same order of magnitude, while c/J_1 is very large. This analysis shows

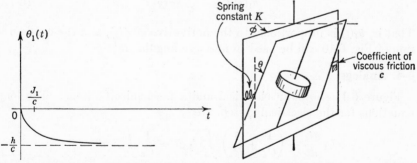

FIG. 6-9. Response of inner gimbal of a floated gyro to a step in outer gimbal angle.   FIG. 6-10. Gyro with spring restraint and viscous damping.

that if the floated gyro is rotated through an angle ϕ, the inner gimbal will quickly settle to an angle $-(h/c)\phi$. The angle θ is therefore a measure of the angle through which the gyro has been rotated.

Rate Gyro. Let us now restrain the inner gimbal of Fig. 6-7 by both a viscous torque (as in the case of the floated gyro) and a spring torque. The new configuration is shown in Fig. 6-10. The moment M_θ is now given by

$$M_\theta = -K\theta - c\dot\theta \tag{6-72}$$

and Eq. (6-67) becomes

$$\begin{bmatrix} J_1 s^2 & -hs \\ hs & J_1 s^2 + cs + K \end{bmatrix} \begin{bmatrix} \bar\phi(s) \\ \bar\theta_1(s) \end{bmatrix} = \begin{bmatrix} \bar M_\phi \\ 0 \end{bmatrix} \tag{6-73}$$

The transforms of the angles are now

$$\bar\theta_1(s) = \frac{\begin{vmatrix} J_1 s^2 & \bar M_\phi \\ hs & 0 \end{vmatrix}}{\begin{vmatrix} J_1 s^2 & -hs \\ hs & J_1 s^2 + cs + K \end{vmatrix}}$$

$$\bar\phi(s) = \frac{\begin{vmatrix} \bar M_\phi & -hs \\ 0 & J_1 s^2 + cs + K \end{vmatrix}}{\begin{vmatrix} J_1 s^2 & -hs \\ hs & J_1 s^2 + cs + K \end{vmatrix}} \tag{6-74}$$

The ratio is

$$\frac{\bar\theta_1(s)}{\bar\phi(s)} = \frac{\begin{vmatrix} J_1 s^2 & \bar M_\phi \\ hs & 0 \end{vmatrix}}{\begin{vmatrix} \bar M_\phi & -hs \\ 0 & J_1 s^2 + cs + K \end{vmatrix}} = \frac{-\bar M_\phi hs}{\bar M_\phi(J_1 s^2 + cs + K)} = \frac{-\dfrac{h}{J_1} s}{s^2 + \dfrac{c}{J_1} s + \dfrac{K}{J_1}} \tag{6-75}$$

In practice K/J_1 can be made large, so that an approximate relation between $\bar\theta_1(s)$ and $\bar\phi(s)$ is

$$\bar\theta_1(s) \cong -\frac{h}{K} s\bar\phi(s) \tag{6-76}$$

That is, $\theta_1(t)$ is proportional to the derivative of $\phi(t)$, and the configuration of Fig. 6-10 can be used to measure angular rates.

6-4. Analogs

Figure 6-11 shows an electrical and a mechanical system. The loop equations for the electrical system are

$$L_1 \frac{di_1}{dt} + \left(\frac{1}{C_1} + \frac{1}{C_2}\right) \int_0^t i_1 \, dt - \frac{1}{C_2} \int_0^t i_2 \, dt = 0$$

$$-\frac{1}{C_2} \int_0^t i_1 \, dt + L_2 \frac{di_2}{dt} + Ri_2 + \frac{1}{C_2} \int_0^t i_2 \, dt = 0 \tag{6-77}$$

In terms of charge q ($i = dq/dt$) these are

$$L_1 \frac{d^2 q_1}{dt^2} + \left(\frac{1}{C_1} + \frac{1}{C_2}\right) q_1 - \frac{1}{C_2} q_2 = 0$$

$$-\frac{1}{C_2} q_1 + L_2 \frac{d^2 q_2}{dt^2} + R \frac{dq_2}{dt} + \frac{1}{C_2} q_2 = 0 \tag{6-78}$$

The equations for the mechanical system are [from (6-25)]

$$M_1 \frac{d^2x_1}{dt^2} + (K_1 + K_2)x_1 - K_2x_2 = 0$$

$$-K_2x_1 + M_2 \frac{d^2x_2}{dt^2} + B \frac{dx_2}{dt} + K_2x_2 = 0 \tag{6-79}$$

We see that with some changes in symbols Eqs. (6-78) and (6-79) would be the same. Systems with equations of the same form such as these are called *analogs*. The quantities which occupy corresponding positions

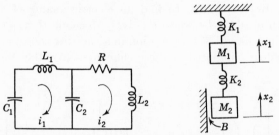

Fig. 6-11. Electrical and mechanical systems which are analogs.

in the two equations are called *analogous quantities*, with the relation indicated by the symbol $\sim$. In this case these are

Charge $q \sim x$　　displacement
Voltage $v \sim f$　　force
Capacitance $C \sim 1/K$　inverse spring constant (compliance)
Inductance $L \sim M$　　mass
Resistance $R \sim B$　　coefficient of viscous friction

From the definition of an analog, we see that the dual networks of Sec. 4-7 are also analogs. In fact, a second analog of the mechanical system in Fig. 6-11 can be formed from the dual of its electric analog shown in Fig. 6-12. The equations are

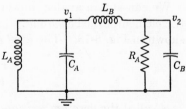

Fig. 6-12. Another analog of the mechanical system of Fig. 6-11.

$$C_A \frac{dv_1}{dt} + \left(\frac{1}{L_A} + \frac{1}{L_B}\right) \int_0^t v_1 \, dt - \frac{1}{L_B} \int_0^t v_2 \, dt = 0$$

$$-\frac{1}{L_B} \int_0^t v_1 \, dt + C_B \frac{dv_2}{dt} + \frac{1}{R_A} v_2 + \frac{1}{L_B} \int_0^t v_2 \, dt = 0 \tag{6-80}$$

In this case the analogous quantities are

$$
\begin{array}{rl}
\text{Voltage} & v \sim \dot{x} \quad \text{velocity} \\
\text{Reciprocal inductance} & 1/L \sim K \quad \text{spring constant} \\
\text{Capacitance} & C \sim M \quad \text{mass} \\
\text{Conductance} \ 1/R \sim B & \text{coefficient of viscous friction} \\
\text{Current} & i \sim f \quad \text{force}
\end{array}
$$

One utility of analogs lies in the fact that electrical systems are often easier to study experimentally than mechanical systems. A resistance, for example, can be changed more easily than a coefficient of friction.

It is not always possible to find an electric analog of a mechanical system which uses Rs, Ls, and Cs only. Mechanical systems with non-linear equations like (6-33) for motion in a rotating coordinate system do not have analogs composed of these elements. Another kind of system

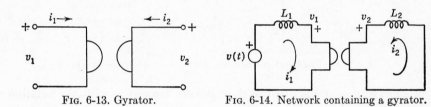

FIG. 6-13. Gyrator.   FIG. 6-14. Network containing a gyrator.

without an RLC analog is represented by Eqs. (6-61). These equations are linear, but the off-diagonal terms corresponding to z_{12} and z_{21} (see Sec. 4-3) are not equal:

$$
\begin{bmatrix} J_1 s^2 & -hs \\ hs & J_1 s^2 \end{bmatrix} \begin{bmatrix} \bar{\phi}(s) \\ \bar{\theta}_1(s) \end{bmatrix} = \begin{bmatrix} \bar{M}_\phi(s) \\ 0 \end{bmatrix} \tag{6-81}
$$

We can get an analog, however, by introducing another kind of electric element, called a *gyrator*.[1] This element is represented by the symbol shown in Fig. 6-13. The *ideal gyrator* is defined by the equations

$$
\begin{aligned}
v_1 &= -R i_2 \\
v_2 &= R i_1
\end{aligned} \tag{6-82}
$$

R is called the *gyrating resistance*.

In matrix notation the equations are

$$
\begin{bmatrix} 0 & -R \\ R & 0 \end{bmatrix} \begin{bmatrix} i_1 \\ i_2 \end{bmatrix} = \begin{bmatrix} v_1 \\ v_2 \end{bmatrix} \tag{6-83}
$$

[1] B. D. H. Tellegen, The Gyrator, A New Electric Network Element, *Philips Research Rept.*, **3**:81–101 (April, 1948).

As defined, the gyrator contains no sources and dissipates no energy. Elements having these characteristics can be constructed.[1]

Let us now consider the electrical network shown in Fig. 6-14. The loop equations are

$$L_1 \frac{di_1}{dt} + v_1 = v(t)$$
$$L_2 \frac{di_2}{dt} + v_2 = 0$$

(6-84)

and for the gyrator

$$v_1 = -Ri_2$$
$$v_2 = Ri_1$$

(6-85)

Combining these, we have

$$L_1 \frac{di_1}{dt} - Ri_2 = v(t)$$
$$L_2 \frac{di_2}{dt} + Ri_1 = 0$$

(6-86)

In terms of charge q, these are

$$L_1 \frac{d^2q_1}{dt^2} - R \frac{dq_2}{dt} = v(t)$$
$$L_2 \frac{d^2q_2}{dt^2} + R \frac{dq_1}{dt} = 0$$

(6-87)

Now we transform with $\mathcal{L}[q(t)] \triangleq \bar{q}(s)$

$$\begin{bmatrix} L_1 s^2 & -Rs \\ Rs & L_2 s^2 \end{bmatrix} \begin{bmatrix} \bar{q}_1(s) \\ \bar{q}_2(s) \end{bmatrix} = \begin{bmatrix} \bar{v}(s) \\ 0 \end{bmatrix}$$

(6-88)

From a comparison with (6-81) we conclude that the network of Fig. 6-14 is an analog of the gyroscope of Fig. 6-7 for small angles with

Inductance $L_1, L_2 \sim J_1$	moment of inertia	
Gyrating resistance $R \sim h$	angular momentum	
Charge $q_1, q_2 \sim \phi, \theta_1$	gimbal angle	
Voltage $v \sim M$	moment	

6-5. Beams

In Sec. 2-2 we wrote down the equation for deflection of a beam. Here we shall show how the equation is derived and apply it to specific examples. As usual, our aim is to keep the derivation concise and emphasize $\mathcal{L}$-transform solutions.[2]

[1] C. L. Hogan, The Elements of Nonreciprocal Microwave Devices, *Proc. IRE*, **44**:1345–1368 (October, 1956).

[2] A good reference on beams is S. Timoshenko and G. H. McCullough, "Elements of Strength of Materials," 3d ed., chaps. 4 and 6, D. Van Nostrand Company, Inc., Princeton, N.J., 1940.

Our problem will be to find the deflection $y(x)$ of a beam due to some general loading $w(x)$ as shown in Fig. 6-15. The loading $w(x)$ has dimensions of force/unit length.

The positive directions of deflection and loading are shown in Fig. 6-15. Notice that x is the independent variable and that in these static problems time does not enter.

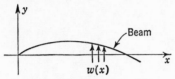

FIG. 6-15. Loaded beam.

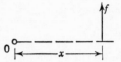

FIG. 6-16. Definition of moment.

Moment, Bending, and Shear. The moment of a force about a point is defined as the product of the distance from point to force and the normal component of the force. The moment of f about the point 0 in Fig. 6-16 is xf. The sign of the moment is positive if it is produced by an upward force.

Shearing force V at a point is equal to the algebraic sum of all forces on either side of the point. The sign can be determined from Fig. 6-17, which shows the type of deformation which could be caused by shearing forces.

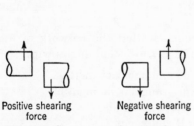

FIG. 6-17. Sign conventions for shear.

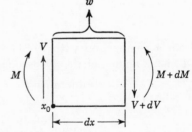

FIG. 6-18. Small element of a loaded beam.

Let us consider the forces and moments acting on a small element of a beam shown in Fig. 6-18. The loading is assumed constant, since dx is small. Writing the equations of equilibrium, we have for the condition that the moments must vanish about x_0

$$(M + dM) - M - (V + dV)\,dx + \int_0^{(dx)} xw\,dx = 0$$

$$dM - V\,dx - dV\,dx + w\,\frac{(dx)^2}{2} = 0 \tag{6-89}$$

Dropping products of differentials, [if dx is small, $(dx)^2$ can be neglected] we have

$$dM - V\,dx = 0$$

$$V = \frac{dM}{dx} \tag{6-90}$$

The equilibrium condition that the sum of forces must vanish is written

$$w\,dx + V - (V + dV) = 0$$

$$w = \frac{dV}{dx} \tag{6-91}$$

Combining (6-90) and (6-91), we have

$$w = \frac{d^2M}{dx^2} \tag{6-92}$$

Derivation of the Beam Equation. We shall now derive a relation between bending moment M and deflection y. We begin with Hooke's

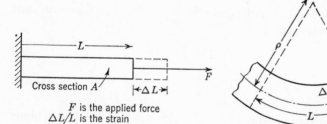

F is the applied force
$\Delta L/L$ is the strain

Fig. 6-19. Beam under tension.

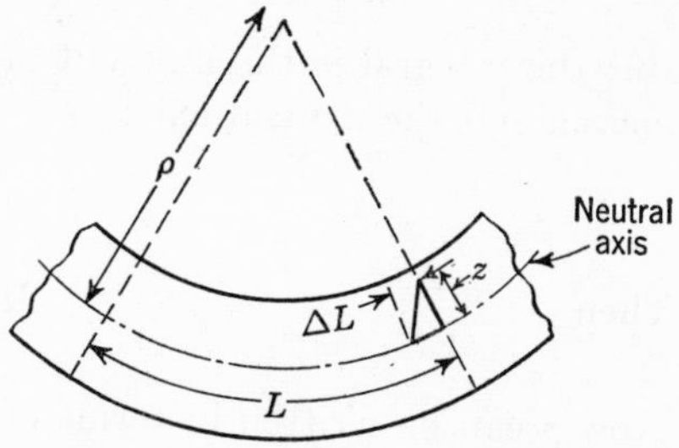

Fig. 6-20. Segment of a beam deformed by bending.

law: Stress is proportional to strain. In terms of the quantities in Fig. 6-19, Hooke's law states that

$$E\frac{\Delta L}{L} = \frac{F}{A} \tag{6-93}$$

where E is a constant of proportionality called *modulus of elasticity.*

Let us now consider an element of a beam which is deformed by bending as shown in Fig. 6-20. The radius of curvature is ρ. The neutral axis is the projection of a surface along which no strain occurs during bending. By similar triangles we can write

$$\frac{\Delta L}{z} = \frac{L}{\rho}$$

$$\frac{\Delta L}{L} = \frac{z}{\rho} \tag{6-94}$$

Let us now examine the forces acting on a cross section of the beam.

Figure 6-21 shows a force dF acting at an element of area dA. From (6-93) and (6-94) we write for a fiber of the beam of cross section dA

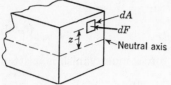

FIG. 6-21. Cross section of the beam.

$$E \frac{\Delta L}{L} = E \frac{z}{\rho} = \frac{dF}{dA} \qquad (6\text{-}95)$$

or

$$dF = \frac{E}{\rho} z \, dA \qquad (6\text{-}96)$$

Now the moment of the force dF about the neutral axis is

$$dM = z \, dF = \frac{E}{\rho} z^2 \, dA \qquad (6\text{-}97)$$

The total moment is found by integrating over the whole area A:

$$M = \int_A dM = \frac{E}{\rho} \int_A z^2 \, dA \qquad (6\text{-}98)$$

But the integral is the moment of inertia of the cross section with unit mass about the neutral axis:

$$\int_A z^2 \, dA \triangleq I \qquad (6\text{-}99)$$

Then

$$M = \frac{EI}{\rho} \qquad (6\text{-}100)$$

Now ρ can be written in terms of beam deflection y and distance along the beam x:[1]

$$\frac{1}{\rho} = \frac{d^2y/dx^2}{[1 + (dy/dx)^2]^{3/2}} \simeq \frac{d^2y}{dx^2} \qquad \text{for } \frac{dy}{dx} \text{ small} \qquad (6\text{-}101)$$

The result (6-100) now becomes

$$M = EI \frac{d^2y}{dx^2} \qquad (6\text{-}102)$$

and when this is combined with (6-92), we have finally

$$EI \frac{d^4y}{dx^4} = w(x) \qquad (6\text{-}103)$$

We shall now illustrate the application of this equation to examples.

[1] The expression for radius of curvature is derived in most calculus books. See, for example, G. E. F. Sherwood and A. E. Taylor, "Calculus," p. 163, Prentice-Hall, Inc., Englewood Cliffs, N.J., 1942.

Equation (6-103) was transformed in Sec. 2-2. We repeat the result here, letting $Y(s) \triangleq \mathcal{L}[y(x)]$:

$$Y(s) = \frac{y(0)}{s} + \frac{y'(0)}{s^2} + \frac{y''(0)}{s^3} + \frac{y'''(0)}{s^4} + \frac{W(s)}{EIs^4} \qquad (6\text{-}104)$$

where $y(0)$ = deflection at $x = 0$
$\quad y'(0)$ = slope at $x = 0$
$\quad y''(0) = (1/EI) \cdot$ bending moment at $x = 0$ $\qquad (6\text{-}105)$
$\quad y'''(0) = (1/EI) \cdot$ shear at $x = 0$
$\quad W(s) = \mathcal{L}[w(x)]$

Example 1. Figure 6-22 shows a uniformly loaded cantilever beam. The conditions are

$$
\begin{aligned}
&y(0) = 0 \\
&y'(0) = 0 \\
&y''(0) = -\left(\frac{1}{EI}\right) \int_0^L w_0 x \, dx = -\frac{(1/EI)w_0 L^2}{2} \\
&y'''(0) = \frac{w_0 L}{EI} \\
&w(x) = -w_0 u(x)
\end{aligned}
\qquad (6\text{-}106)
$$

The transformed deflection is

$$Y(s) = -\frac{w_0 L^2}{2EIs^3} + \frac{w_0 L}{EIs^4} - \frac{w_0}{EIs^5} \qquad (6\text{-}107)$$

which gives for $y(x)$

$$y(x) = -\frac{w_0 L^2}{4EI} x^2 + \frac{w_0 L}{6EI} x^3 - \frac{w_0}{24EI} x^4 \qquad 0 \le x \le L \quad (6\text{-}108)$$

It is not necessary to reduce the loading to zero for $x > L$ in defining $w(x)$ in problems of this type.

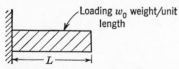

FIG. 6-22. Uniformly loaded cantilever beam.

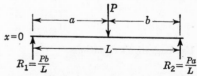

FIG. 6-23. Simple beam with concentrated load.

Example 2. The conditions are not always given at $x = 0$ as in the first example. To show how this situation can be handled, we consider the beam shown in Fig. 6-23. R_1 and R_2 are reactions at the supports. The load has weight P. The conditions of the problem are

$$y(0) = y(L) = 0$$
$$y''(0) = 0 \quad \text{(no moment at point support)}$$
$$y'''(0) = \frac{R_1}{EI} \tag{6-109}$$
$$w(x) = -P\delta(x - a)$$

The slope at $x = 0$ cannot be determined yet.   The transformed equation is

$$Y(s) = \frac{y'(0)}{s^2} + \frac{R_1}{EIs^4} - \frac{Pe^{-as}}{s^4 EI} \tag{6-110}$$

$y'(0)$ is as yet unknown.   Taking the inverse transform,[1]

$$y(x) = y'(0)x + \frac{R_1}{6EI} x^3 - \frac{P}{6EI} (x - a)^3 u(x - a) \tag{6-111}$$

Using the condition that $y(L) = 0$, we write

$$y(L) = 0 = y'(0)L + \frac{R_1}{6EI} L^3 - \frac{P}{6EI} (L - a)^3$$
$$= y'(0)L + \frac{PbL^2}{6EI} - \frac{Pb^3}{6EI} \tag{6-112}$$

Solving for $y'(0)$, we have

$$y'(0) = -\frac{PbL}{6EI} + \frac{Pb^3}{6EIL} \tag{6-113}$$

which when substituted in (6-111) completes the solution:

$$y(x) = \frac{Pb}{6EIL} \left[(b^2 - L^2)x + x^3 - \frac{L}{b} (x - a)^3 u(x - a) \right] \tag{6-114}$$

PROBLEMS

6-1. Show that the equations of motion for the system shown are

$$M_1\ddot{x}_1 + (K_1 + K_2)x_1 - K_2 x_2 = 0$$
$$M_2\ddot{x}_2 + (K_2 + K_3)x_2 - K_2 x_1 = 0$$

Use Lagrange's equations.

PROB. 6-1

[1] We make use of a relation we shall prove later (Sec. 8-2):

$$\mathcal{L}^{-1}[e^{-as}F(s)] = f(t - a)u(t - a) \qquad a \geq 0$$

6-2. Show from Lagrange's equations that the equation for the Atwood's machine shown is

$$\ddot{x} = \frac{M_1 - M_2}{M_1 + M_2} g$$

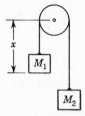

PROB. 6-2

6-3. Find the displacement $x_2(t)$ in Prob. 6-1 due to an initial displacement of $x_1(0) = x_0$, $x_2(0) = 0$. Assume that all spring constants are K and both masses M.

6-4. Write the equations of motion for the system shown.

Ans. $\ddot{x}_1(M_1 + M_2 + M_3) - \ddot{x}_2(M_2 - M_3) = g(M_3 + M_2 - M_1)$
$-\ddot{x}_1(M_2 - M_3) + \ddot{x}_2(M_3 + M_2) = g(M_3 - M_2)$

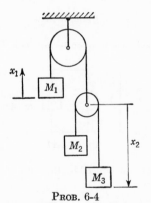

PROB. 6-4

6-5. A ball of mass m drops through a liquid of constant viscous damping coefficient b. The upward buoyant force increases linearly with depth and has magnitude Ax at depth x. Assume that the ball starts at the surface with zero initial velocity. Find $x(t)$ with $b = 4$, $A = 4$, $m = 1$.

6-6. Find electrical analogs for systems in Probs. 6-1, 6-2, 6-4, and 6-5.

6-7. Find two electrical analogs for the system in Fig. 6-2.

6-8. Write the transformed equation of the continuously loaded cantilever beam shown.

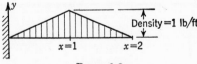

PROB. 6-8

6-9. Find the deflection of the cantilever beam shown.

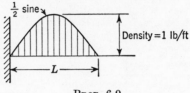

PROB. 6-9

6-10. Find the equations of motion for the double pendulum shown.

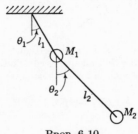

PROB. 6-10

6-11. Show that the impedance between terminals is R^2/Z_L.

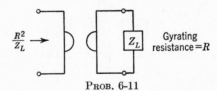

PROB. 6-11

6-12. Show that the power introduced by the voltage source is all dissipated in the load R_L.

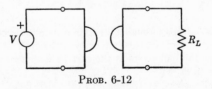

PROB. 6-12

6-13. Find an electrical analog for the rate gyro of Fig. 6-10.

CHAPTER 7

INVERSE TRANSFORMS

The solutions of physical problems like the ones discussed in preceding chapters involve finding the inverse $\mathcal{L}$ transforms of functions like the one occurring in (4-17):

$$F(s) = \frac{-s^4 - s^3 - 2s^2 + 2s + 2}{s(s^4 + 6s^3 + 10s^2 + 8s + 4)} \tag{7-1}$$

which is the transform of the current in a network. Most of the transforms in this book will be *rational fractions* like (7-1). That is, they will be ratios of polynomials in s. Rational fractions arise in problems which involve linear, ordinary, constant-coefficient differential equations. Other kinds of transforms occur, especially in some partial differential equations and in nonconstant-coefficient equations, but their inversion requires techniques beyond those we shall take up.

In this chapter we shall discuss ways for finding the time functions associated with $\mathcal{L}$ transforms which are rational fractions.

7-1. Transform Tables

It goes without saying that when a function of s appears in a transform table, like the one in the back of this book, our problem is solved. It is usually true, however, that a given problem will not yield transforms in just the tabulated form. For example, it would be unlikely that (7-1) would appear in even a very large table of transforms. Before we get (7-1) in shape for inversion, let us discuss a property of $\mathcal{L}$ transforms which is useful in extending a transform table.

If a transform contains a parameter, then we are free to give this parameter any value we like. For example, starting with the pair

$$\frac{1}{(s + \alpha)(s + \gamma)} \qquad \bigg| \qquad \frac{e^{-\alpha t} - e^{-\gamma t}}{\gamma - \alpha} \tag{7-2}$$

we can let $\gamma \to 0$, in which case we have a new pair:

$$\frac{1}{s(s + \alpha)} \qquad \bigg| \qquad \frac{1}{\alpha}(1 - e^{-\alpha t}) \tag{7-3}$$

91

If we let $\alpha \to \gamma$ we will get an indeterminate form, $0/0$, on the right, but this can be resolved by using L'Hospital's rule—differentiating numerator and denominator, in this case with respect to α:

$$\frac{1}{(s + \gamma)^2} \qquad \Big| \qquad \lim_{\alpha \to \gamma} \frac{e^{-\alpha t} - e^{-\gamma t}}{\gamma - \alpha} = \frac{te^{-\alpha t}}{1} \Big|_{\alpha = \gamma} = te^{-\gamma t} \qquad (7\text{-}4)$$

Tables of $\mathcal{L}$ transforms are available from a number of sources. In addition to the table of rational fractions in Appendix B of this book, extensive transform tables are given by Gardner and Barnes[1] and Nixon.[2] A large table of transforms of all kinds is a part of the Bateman Manuscript Project.[3]

7-2. Partial Fractions

Let us return to the preparation of a rational fraction in s for inversion. There are two steps which should be taken first:

1. *The rational fraction should be made a proper fraction.* That is, the degree of the numerator should be made less than that of the denominator. Suppose we have

$$F_1(s) = \frac{s^4 + 8s^3 + 25s^2 + 31s + 15}{s^3 + 6s^2 + 11s + 6} \qquad (7\text{-}5)$$

We simply perform a long division:

$$
\begin{array}{l}
s^4 + 8s^3 + 25s^2 + 31s + 15 \;\underline{\big|\, s^3 + 6s^2 + 11s + 6} \\
\underline{s^4 + 6s^3 + 11s^2 + 6s} \qquad\quad s + 2 \\
\quad\; 2s^3 + 14s^2 + 25s + 15 \\
\quad\; \underline{2s^3 + 12s^2 + 22s + 12} \\
\qquad\qquad 2s^2 + 3s + 3
\end{array}
\qquad (7\text{-}5a)
$$

so that we can write

$$F_1(s) = s + 2 + \frac{2s^2 + 3s + 3}{s^3 + 6s^2 + 11s + 6} \qquad (7\text{-}6)$$

We now have terms which will yield impulse functions plus a proper rational fraction. The division has separated out the impulses:

$$\mathcal{L}^{-1}[F_1(s)] = \delta'_+(t) + 2\delta_+(t) + \mathcal{L}^{-1}\left[\frac{2s^2 + 3s + 3}{s^3 + 6s^2 + 11s + 6}\right] \qquad (7\text{-}7)$$

[1] M. F. Gardner and J. L. Barnes, "Transients in Linear Systems," John Wiley & Sons, Inc., New York, 1942.

[2] F. E. Nixon, "Principles of Automatic Control," pp. 371–399, Prentice-Hall, Inc., Englewood Cliffs, N.J., 1953.

[3] A. Erdélyi, "Tables of Integral Transforms," vol. 1, McGraw-Hill Book Company, Inc., New York, 1954.

2. *The denominator of the rational fraction should be factored.* Our objective is to expand rational fractions into partial fractions with terms which can be looked up in the table separately. Such an expansion for the rational fraction in (7-7) is

$$\frac{2s^2 + 3s + 3}{s^3 + 6s^2 + 11s + 6} = \frac{2s^2 + 3s + 3}{(s + 1)(s + 2)(s + 3)} = \frac{1}{s + 1} - \frac{5}{s + 2} + \frac{6}{s + 3}$$
(7-8)

so that we have the result

$$\mathcal{L}^{-1}[F_1(s)] = \delta'_+(t) + 2\delta_+(t) + e^{-t} - 5e^{-2t} + 6e^{-3t} \qquad (7-9)$$

The coefficients of the terms on the right of (7-8) could be found by writing

$$\frac{2s^2 + 3s + 3}{(s + 1)(s + 2)(s + 3)} = \frac{A}{s + 1} + \frac{B}{s + 2} + \frac{C}{s + 3}$$

$$= \frac{s^2(A + B + C) + s(5A + 4B + 3C) + (6A + 3B + 2C)}{(s + 1)(s + 2)(s + 3)} \qquad (7-10)$$

Equating coefficients of powers of s yields a set of equations

$$\begin{aligned} A + B + C &= 2 \\ 5A + 4B + 3C &= 3 \\ 6A + 3B + 2C &= 3 \end{aligned} \qquad (7-11)$$

which has the solution $A = 1$, $B = -5$, $C = 6$. We shall next consider a more direct way of obtaining the coefficients.

The general rational fraction will contain second- or higher-order poles. We shall expand such functions into partial fractions as in the example:

$$F_2(s) = \frac{2s^2 + 3s + 3}{(s + 1)(s + 3)^3} = \frac{A}{s + 1} + \frac{B_3}{(s + 3)^3} + \frac{B_2}{(s + 3)^2} + \frac{B_1}{s + 3} \qquad (7-12)$$

That is, there will be a number of terms for each pole equal to the order of the pole. Referring now to (7-12) we see that

$$[(s + 1)F_2(s)]_{s=-1}$$

$$= \left[A + \frac{(s + 1)B_3}{(s + 3)^3} + \frac{(s + 1)B_2}{(s + 3)^2} + \frac{(s + 1)B_1}{s + 3} \right]_{s=-1} = A \qquad (7-13)$$

and similarly

$$[(s + 3)^3 F_2(s)]_{s=-3} = B_3 \qquad (7-14)$$

The other coefficients can be obtained by differentiating:

$$\left[\frac{d}{ds}(s+3)^3 F_2(s)\right]_{s=-3} = \left\{A\left[\frac{3(s+3)^2(s+1) - (s+3)^3}{(s+1)^2}\right]\right.$$
$$\left.+ \frac{d}{ds}B_3 + B_2 + 2(s+3)B_1\right\}_{s=-3} = B_2 \quad (7\text{-}15)$$

and similarly

$$\left[\frac{d^2}{ds^2}(s+3)^3 F_2(s)\right]_{s=-3} = 2B_1 \quad (7\text{-}16)$$

We have shown in the example how the partial-fraction coefficients can be obtained from the original function.   The results can be summarized:
In the expansion of $F(s)$, a proper rational fraction

$$F(s) = \frac{G(s)}{(s-s_1)^{m_1}(s-s_2)^{m_2}\cdots(s-s_k)^{m_k}}$$
$$= \frac{A_{m_1}}{(s-s_1)^{m_1}} + \frac{A_{m_1-1}}{(s-s_1)^{m_1-1}} + \cdots + \frac{A_1}{(s-s_1)}$$
$$+ \frac{B_{m_2}}{(s-s_2)^{m_2}} + \cdots \quad (7\text{-}17)$$

The coefficients are given by

$$A_{m_1} = [(s-s_1)^{m_1}F(s)]_{s=s_1}$$
$$A_{m_1-1} = \left[\frac{d}{ds}(s-s_1)^{m_1}F(s)\right]_{s=s_1}$$
$$A_{m_1-2} = \left[\frac{1}{2!}\frac{d^2}{ds^2}(s-s_1)^{m_1}F(s)\right]_{s=s_1} \quad (7\text{-}18)$$
$$\cdots\cdots\cdots\cdots\cdots\cdots\cdots$$
$$A_1 = \left[\frac{1}{(m_1-1)!}\frac{d^{m_1-1}}{ds^{m_1-1}}(s-s_1)^{m_1}F(s)\right]_{s=s_1}$$
$$\cdots\cdots\cdots\cdots\cdots\cdots\cdots$$

To illustrate, we consider the following examples:

Example 1.   From (7-10), we have

$$\frac{2s^2 + 3s + 3}{(s+1)(s+2)(s+3)} = \frac{A}{s+1} + \frac{B}{s+2} + \frac{C}{s+3} \quad (7\text{-}19)$$

Then from (7-18)

$$A = \left[\frac{2s^2 + 3s + 3}{(s+2)(s+3)}\right]_{s=-1} = \frac{2}{2} = 1$$
$$B = \left[\frac{2s^2 + 3s + 3}{(s+1)(s+3)}\right]_{s=-2} = \frac{5}{-1} = -5 \quad (7\text{-}20)$$
$$C = \left[\frac{2s^2 + 3s + 3}{(s+1)(s+2)}\right]_{s=-3} = \frac{12}{2} = 6$$

Example 2. For the example (7-12) we have

$$A = \left[\frac{2s^2 + 3s + 3}{(s + 3)^3}\right]_{s=-1} = \frac{1}{4}$$

$$B_3 = \left[\frac{2s^2 + 3s + 3}{s + 1}\right]_{s=-3} = -6$$

$$B_2 = \left[\frac{d}{ds}\frac{2s^2 + 3s + 3}{s + 1}\right]_{s=-3} = \left[\frac{2s(s + 2)}{(s + 1)^2}\right]_{s=-3} = \frac{3}{2}$$ (7-21)

$$B_1 = \left[\frac{1}{2}\frac{d^2}{ds^2}\frac{2s^2 + 3s + 3}{s + 1}\right]_{s=-3} = \left[\frac{2}{(s + 1)^3}\right]_{s=-3} = -\frac{1}{4}$$

7-3. Complex Roots

The results of the previous section are sufficient to expand a general rational fraction into partial fractions.   However, when the rational fraction has a denominator containing complex roots, it may be convenient to modify the expansion.   If we have such a rational fraction, $F(s) = F_1(s)/[(s + \alpha)^2 + \beta^2]$ [where $F_1(s)$ is itself a rational fraction], we write

$$F(s) = F_1(s)\frac{1}{(s + \alpha)^2 + \beta^2} = \frac{A}{s + (\alpha - j\beta)} + \frac{B}{s + (\alpha + j\beta)} + \cdots$$
$$\text{(7-22)}$$

It is sometimes convenient to add the first two terms on the right of (7-22) so that a quadratic term appears in the expansion.

$$F(s) = F_1(s)\frac{1}{[(s + \alpha)^2 + \beta^2]} = \frac{Cs + D}{[(s + \alpha)^2 + \beta^2]} + \cdots \quad \text{(7-23)}$$

We shall now find an expression for the numerator of the first term on the right of (7-23).

We can argue that since $F(s)$ has real coefficients, the first two terms on the right of (7-22) must be complex conjugates so that their sum will have real coefficients.   We can write the sum as twice the real part of either term:[1]

$$F(s) = F_1(s)\frac{1}{(s + \alpha)^2 + \beta^2} = 2\,\mathrm{Re}\left[\frac{A}{s + \alpha - j\beta}\right] + \cdots$$
$$= \frac{2\,\mathrm{Re}\,[A(s + \alpha + j\beta)]}{[(s + \alpha)^2 + \beta^2]} + \cdots \quad \text{(7-24)}$$

Now from (7-18) the coefficient A is given by

$$A = [(s + \alpha - j\beta)F(s)]_{s=-\alpha+j\beta}$$
$$= \frac{F_1(-\alpha + j\beta)}{-\alpha + j\beta + \alpha + j\beta} = \frac{F_1(-\alpha + j\beta)}{2j\beta} \quad \text{(7-25)}$$

[1] See Appendix A, Eq. (A-10).

so that the expansion (7-23) becomes[1]

$$F(s) = F_1(s) \frac{1}{[(s + \alpha)^2 + \beta^2]} = \text{Re} \left[\frac{F_1(-\alpha + j\beta)}{j\beta(s + \alpha - j\beta)} \right] + \cdots$$

$$= \frac{\text{Re} \left[\frac{(s + \alpha + j\beta)F_1(-\alpha + j\beta)}{j\beta} \right]}{[(s + \alpha)^2 + \beta^2]} + \cdots$$

(7-26)

This result will be used in Sec. 9-4 when we discuss the response of a system to a sine wave.

We shall illustrate with two examples.

Example 1. Let us expand the function

$$F(s) = \frac{s + 3}{s^3 + 3s^2 + 6s + 4} = \frac{s + 3}{[(s + 1)^2 + 3](s + 1)} \tag{7-27}$$

In this case [referring to (7-26)]

$$F_1(s) = \frac{s + 3}{s + 1} \qquad \alpha = 1 \qquad \beta = \sqrt{3} \tag{7-28}$$

Then

$$\text{Re} \left[\frac{(s + \alpha + j\beta)F_1(-\alpha + j\beta)}{j\beta} \right] = \text{Re} \left[\frac{(s + 1 + j\sqrt{3})(2 + j\sqrt{3})}{(j\sqrt{3})(j\sqrt{3})} \right]$$

$$= \frac{2(s + 1) - 3}{-3} = -\frac{2}{3}s + \frac{1}{3} \tag{7-29}$$

and the expansion is

$$\frac{s + 3}{[(s + 1)^2 + 3](s + 1)} = \frac{2}{3} \frac{1}{s + 1} - \frac{2}{3} \frac{s - \frac{1}{2}}{[(s + 1)^2 + 3]} \tag{7-30}$$

The corresponding time function is, from the table in Appendix B,

$$\mathcal{L}^{-1} \left[\frac{s + 3}{[(s + 1)^2 + 3](s + 1)} \right] = \frac{2}{3} \mathcal{L}^{-1} \left[\frac{1}{s + 1} \right] - \frac{2}{3} \mathcal{L}^{-1} \left[\frac{s - \frac{1}{2}}{(s + 1)^2 + 3} \right]$$

$$= \frac{2}{3} e^{-t} - \frac{2}{3} \frac{1}{\sqrt{3}} \frac{\sqrt{21}}{2} e^{-t} \sin \left(\sqrt{3}\, t + \pi - \tan^{-1} \frac{2}{\sqrt{3}} \right)$$

$$= \frac{2}{3} e^{-t} \left[1 + \frac{\sqrt{7}}{2} \sin \left(\sqrt{3}t - \tan^{-1} \frac{2}{\sqrt{3}} \right) \right] \tag{7-31}$$

Example 2. Let us return to (7-1). This function is already a proper fraction, but its denominator must be factored. The factoring problem

---

[1] In taking the real part in (7-26) s is treated as a real parameter.

is one of numerical analysis and will not be discussed here.[1] Let us assume that the numerical work has been done and write

$$\frac{-s^4 - s^3 - 2s^2 + 2s + 2}{s(s^4 + 6s^3 + 10s^2 + 8s + 4)}$$

$$= \frac{-s^4 - s^3 - 2s^2 + 2s + 2}{s[(s + 0.382)^2 + 0.786^2](s + 3.890)(s + 1.346)}$$

$$= \frac{A}{s} + \frac{Bs + C}{[(s + 0.382)^2 + 0.786^2]} + \frac{D}{s + 3.890} + \frac{E}{s + 1.346} \quad (7\text{-}32)$$

Using the rules for determination of coefficients (7-18) and (7-26), we obtain

$$A = 0.5 \qquad\qquad\qquad\qquad\qquad\qquad\qquad (7\text{-}33)$$

$$Bs + C = \text{Re} \left\{ \frac{(s + 0.382 + j0.786)}{j0.786} \right.$$

$$\left. \left[\frac{-s^4 - s^3 - 2s^2 + 2s + 2}{s(s + 3.890)(s + 1.346)} \right]_{s=-0.382+j0.786} \right\}$$

$$= \text{Re} \left[\frac{(s + 0.382) + j0.786}{j0.786} (-0.289 - j0.677) \right]$$

$$= \frac{-(s + 0.382)0.677 - (0.786)(0.289)}{0.786}$$

$$= -0.861s - 0.618 \quad (7\text{-}34)$$

The other coefficients are found from (7-18):

$$D = -1.645 \qquad E = 0.974 \qquad\qquad (7\text{-}35)$$

We have finally

$$\frac{-s^4 - s^3 - 2s^2 + 2s + 2}{s(s^4 + 6s^3 + 10s^2 + 8s + 4)} = \frac{0.5}{s} - \frac{0.861(s + 0.718)}{(s + 0.382)^2 + 0.786^2}$$

$$- \frac{1.645}{s + 3.890} + \frac{0.974}{s + 1.346} \quad (7\text{-}36)$$

The labor involved in reducing a problem like this one to partial fractions can be considerable. In Sec. 7-5 we shall discuss a means for getting the general behavior of the transient response without extensive numerical analysis. In the next chapter we shall see that the initial and final behavior can be determined without even factoring.

---

[1] A simple method (when it works) to extract roots, real or complex, is Lin's method described in G. S. Brown and D. P. Campbell, "Principles of Servomechanisms," pp. 89–91, John Wiley & Sons, Inc., New York, 1948. A more reliable but somewhat more complicated technique is Graeffe's root-squaring method. A good description of the latter is given by L. A. Pipes, "Applied Mathematics for Engineers and Physicists," 2d ed., pp. 120–127, McGraw-Hill Book Company, Inc., New York, 1958.

7-4. Residues

A method for finding inverse transforms which is distinct from the partial-fraction method is the method of residues. This is a technique for solving (2-5) when $F(s)$ is a rational fraction. The derivation will be found in Appendix A; here we shall describe the method and illustrate by examples.

Associated with each pole[1] of a function of a complex variable is a number called the *residue*. This number is a particular coefficient in the series expansion of the function near the pole. The method of residues states

If $F(s)$ is a rational fraction, then

$$\mathcal{L}^{-1}[F(s)] = \sum_{\substack{\text{all} \\ \text{poles}}} [\text{residues of } F(s)e^{st}] \tag{7-37}$$

where the residue at an nth-order pole at $s = s_1$ is given by

$$R_{s_1} = \frac{1}{(n-1)!} \left[\frac{d^{n-1}}{ds^{n-1}} (s - s_1)^n F(s) e^{st} \right]_{s=s_1} \tag{7-38}$$

We note that the formula for the residue is similar to that for the coefficient of the first-order term in a partial-fraction expansion. The distinction between (7-38) and the partial-fraction formula (7-18) is the presence of the factor exp (st) in the residue formula.

Example 1. Let us take as an example (7-8) and find

$$\mathcal{L}^{-1}[F(s)] = \mathcal{L}^{-1} \left[\frac{2s^2 + 3s + 3}{(s+1)(s+2)(s+3)} \right] \tag{7-39}$$

The residue of $F(s)e^{st}$ at $s = -1$ is

$$R_{-1} = \left[\frac{2s^2 + 3s + 3}{(s+2)(s+3)} e^{st} \right]_{s=-1} = \frac{2e^{-t}}{2} = e^{-t} \tag{7-40}$$

Similarly,

$$R_{-2} = \left[\frac{2s^2 + 3s + 3}{(s+1)(s+3)} e^{st} \right]_{s=-2} = -5e^{-2t} \tag{7-41}$$

and

$$R_{-3} = \left[\frac{2s^2 + 3s + 3}{(s+1)(s+2)} e^{st} \right]_{s=-3} = \frac{12}{2} e^{-3t} = 6e^{-3t} \tag{7-42}$$

The inverse transform is the sum of the residues R_{-1}, R_{-2}, and R_{-3}.

Example 2. As a second example to illustrate the method when applied to higher-order poles, let us generalize (7-12) and find

$$\mathcal{L}^{-1} \left[\frac{a_2 s^2 + a_1 s + a_0}{(s + \alpha)(s + \gamma)^3} \right] \tag{7-43}$$

---

[1] See Sec. 2-5 for definition of pole.

The residue at the pole at $s = -\alpha$ is

$$R_{-\alpha} = \left[\frac{a_2 s^2 + a_1 s + a_0}{(s+\gamma)^3} e^{st} \right]_{s=-\alpha} = \frac{a_2 \alpha^2 - a_1 \alpha + a_0}{(\gamma - \alpha)^3} e^{-\alpha t} \quad (7\text{-}44)$$

At the other (third-order) pole we have

$$R_{-\gamma} = \left[\frac{1}{2} \frac{d^2}{ds^2} \frac{a_2 s^2 + a_1 s + a_0}{s + \alpha} e^{st} \right]_{s=-\gamma}$$

$$= \frac{1}{2} \left\{ \frac{d}{ds} \left[e^{st} \frac{a_2 s^2 + 2a_2 \alpha s + a_1 \alpha - a_0}{(s+\alpha)^2} \right. \right.$$

$$\left. \left. + te^{st} \frac{a_2 s^3 + (a_1 + a_2 \alpha)s^2 + (a_1 \alpha + a_0)s + a_0 \alpha}{(s+\alpha)^2} \right] \right\}_{s=-\gamma} \quad (7\text{-}45)$$

$$= \left[e^{st} \frac{a_2 \alpha^2 - a_1 \alpha + a_0}{(s+\alpha)^3} + te^{st} \frac{a_2 s^2 + 2a_2 \alpha s + a_1 \alpha - a_0}{(s+\alpha)^2} \right.$$

$$\left. + \frac{1}{2} t^2 e^{st} \frac{a_2 s^2 + a_1 s + a_0}{(s+\alpha)} \right]_{s=-\gamma} \quad (7\text{-}46)$$

$$= \frac{a_2 \alpha^2 - a_1 \alpha + a_0}{(\alpha - \gamma)^3} e^{-\gamma t} + \frac{a_2 \gamma^2 - 2a_2 \alpha \gamma + a_1 \alpha - a_0}{(\alpha - \gamma)^2} te^{-\gamma t}$$

$$+ \frac{a_2 \gamma^2 - a_1 \gamma + a_0}{2(\alpha - \gamma)} t^2 e^{-\gamma t} \quad (7\text{-}47)$$

The amount of algebra required to obtain (7-47) is about the same as needed to get the result by partial fractions, so that from the point of view of labor involved there is not much difference in the methods. We can make the following comparison:

1. The partial-fraction expansion of a rational fraction reduces a complicated transform to the sum of simpler transforms which can be looked up in a table.

The residue method applied to the same transform provides the inverse transform directly.

2. There are n terms associated with an nth-order pole in a partial-fraction expansion. There is only one residue for each pole, regardless of order. Finding the residue for an nth-order pole, however, requires about the same amount of algebra as is required for the partial-fraction terms.

3. The partial-fraction method cannot be applied to a transform like

$$F(s) = \frac{1}{(s+\alpha)^4} \quad (7\text{-}48)$$

because it is already in the form of a term in a partial-fraction expansion and no further reduction is possible. The inverse can be found by residues, however:

$$\mathcal{L}^{-1}\left[\frac{1}{(s+\alpha)^4}\right] = \frac{1}{3!}\left[\frac{d^3}{ds^3}e^{st}\right]_{s=-\alpha} = \frac{t^3}{3!}e^{-\alpha t} \tag{7-49}$$

In summary, partial-fraction expansions are useful in finding inverse transforms when used with a table. They exhibit a transform as a sum of simpler transforms and make predictions about the inverse possible without actually finding the time function as we shall see in the next section. The residue method produces the inverse transform directly without recourse to tables (but without appreciable saving of labor).

7-5. Poles and Zeros

In Sec. 2-5 we defined poles and zeros and stated that the knowledge of the location of poles and zeros of a transform provides information about the corresponding time function.[1] The partial-fraction expansion of a rational fraction is the key to this information, since it exhibits the transform as the sum of terms, each associated with a pole. For example, (7-12) is written

$$\frac{2s^2 + 3s + 3}{(s+1)(s+3)^3} = \frac{1}{4(s+1)} - \frac{6}{(s+3)^3} + \frac{3}{2(s+3)^2} - \frac{1}{4(s+3)} \tag{7-50}$$

Evidently the time function corresponding to (7-50) will be the sum of a term of the form e^{-t} associated with the pole at $s = -1$, and terms of the form t^2e^{-3t}, te^{-3t}, and e^{-3t} from the third-order pole at $s = -3$. We observe also that the two zeros of the left side of (7-50) do not affect the time function in so direct a way.

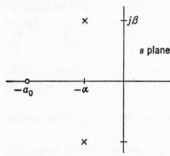

As a second example, let us consider the function with poles and zeros plotted in Fig. 7-1. To within a multiplicative constant the function of s is

FIG. 7-1. s-plane plot of poles and zeros.

$$\frac{s + a_0}{(s+\alpha)^2 + \beta^2} \tag{7-51}$$

and its inverse transform is (from the table in Appendix B)

$$\frac{1}{\beta}[(a_0 - \alpha)^2 + \beta^2]^{1/2}e^{-\alpha t}\sin(\beta t + \psi)$$

$$\psi = \begin{cases} \tan^{-1}\dfrac{\beta}{a_0 - \alpha} & a_0 > \alpha \\[3mm] \tan^{-1}\dfrac{\beta}{a_0 - \alpha} + \pi & a_0 < \alpha \end{cases} \tag{7-52}$$

---

[1] We refer to the inverse transform as a function of time, but it could, of course, be a function of another independent variable like distance.

We see that the time function is a sine wave of frequency β multiplied by exp $(-\alpha t)$. The zero at $-a_0$ does not affect the *time* behavior (frequency, damping) but only the amplitude and phase angle.

To summarize: A rational-fraction transform can be expanded into partial fractions with terms for each pole. The associated time function will be the sum of terms associated with the poles, and the *form* of the time function will be determined by the location of the poles in the s plane:

1. Poles on the negative real axis contribute terms of the form $e^{-\alpha t}$, $te^{-\alpha t}$, etc., depending on the order of the pole.

2. Complex poles in the left half plane contribute damped sine waves.

3. Purely imaginary poles contribute sine waves.

4. Poles with positive real parts contribute terms which grow exponentially with time.

The zeros of the transform contribute to amplitudes and phase angles but do not influence the *form* of the time function.

Usually the first step in analyzing a linear system is to determine in a rough way the response to a "typical" input. If the poles of the transform of this response are known, this rough information can be determined quickly from the locations of the poles in the s plane. For example, poles in the right half plane will mean that the response will grow with time and the system will be unstable. In Chap. 10 we shall outline techniques of feedback-system design which are based on the ideas developed in this section.

7-6. s-Plane Geometry

In the last section we discussed the information that can be obtained about a time function from the position of its poles in the s plane. We saw that frequency and damping or time constant associated with each pole can be determined very simply. We shall now look into the possibility of finding amplitude and phase angle.

If $F(s)$ has a first-order pole at $s = s_1$:

$$F(s) = \frac{(s - s_2)(s - s_4) \cdots (s - s_n)}{(s - s_1)(s - s_3) \cdots (s - s_m)} \tag{7-53}$$

then from (7-38) there will be a term in the corresponding time function of the form

$$f_1(t) = \frac{(s_1 - s_2)(s_1 - s_4) \cdots (s_1 - s_n)}{(s_1 - s_3) \cdots (s_1 - s_m)} e^{s_1 t} \tag{7-54}$$

In general the s_1, s_2, etc., will be complex numbers, so we can write (7-54)

$$f_1(t) = \left| \frac{(s_1 - s_2)(s_1 - s_4) \cdots (s_1 - s_n)}{(s_1 - s_3) \cdots (s_1 - s_m)} \right| e^{s_1 t + j\psi_1}$$

$$= \frac{|s_1 - s_2|\,|s_1 - s_4| \cdots |s_1 - s_n|}{|s_1 - s_3| \cdots |s_1 - s_m|} e^{s_1 t + j\psi_1} \tag{7-55}$$

where ψ_1 is the angle of the coefficient of $e^{s_1 t}$ in (7-54).

Let us suppose now that the poles and zeros of $F(s)$ are as sketched in Fig. 7-2.   Directed line segments have been drawn to represent the numbers s_1 and s_2, and one representing $(s_1 - s_2)$ also.   Now $(s_1 - s_2)$

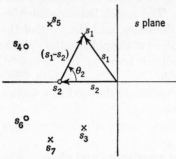

FIG. 7-2. Poles and zeros of $F(s)$.

is a term in the coefficient of (7-54). The magnitude of this term is the length of the line, and the angle is θ_2. The magnitudes and angles of the other terms can be found in the same way. The coefficient of $e^{s_1 t + i\psi_1}$ in (7-55) is simply the product of the lengths from the zeros to s_1 divided by the product of the lengths from the poles to s_1.   The angle ψ_1 is the sum of the angles from the zeros to s_1 minus the sum of angles from the poles to s_1.   Both magnitude and angle of the coefficient of each term can be measured on the s plane with ruler and protractor.

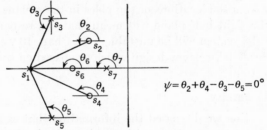

FIG. 7-3. Angle of real pole coefficient.

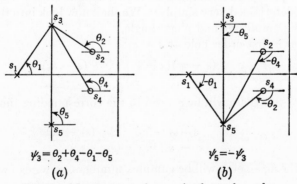

FIG. 7-4. Measurements for a pair of complex poles.

When the pole or zero is complex, as in our example above, it will always be accompanied by its complex conjugate.   This follows from the fact that the poles and zeros are the roots of numerator and denominator of a rational fraction with real coefficients. If in (7-55) s_1 is real, the angle due to the complex poles and zeros is zero, since these are symmetric about the real axis as shown in Fig. 7-3.

The angle due to real poles and zeros is an even or odd multiple of π, making the coefficient positive if there is an even number of poles and zeros to the right of the pole in question or negative if the number is odd.

When the coefficient angles for a pair of complex-pole terms are measured, each will be the negative of the other as shown in Fig. 7-4. The magnitudes will be the same. If $s_3 = -\alpha + j\beta$ and $s_5 = -\alpha - j\beta$, then the sum of corresponding time terms will be, from (7-55),

$$
\begin{aligned}
f_3(t) + f_5(t) &= \frac{|s_3 - s_2|\,|s_3 - s_4|\;\cdots\;|s_3 - s_n|}{|s_3 - s_1|\,|s_3 - s_5|\;\cdots\;|s_3 - s_m|}\,[e^{s_3t+j\psi_3} + e^{s_5t-j\psi_3}] \\
&= M_3 e^{-\alpha t}[e^{j(\beta t+\psi_3)} + e^{-j(\beta t+\psi_3)}] \\
&= 2M_3 e^{-\alpha t}\cos{(\beta t + \psi_3)}
\end{aligned}
\tag{7-56}
$$

where M_3 is the magnitude associated with the pole at s_3:

$$
M_3 \triangleq \frac{|s_3 - s_2|\,|s_3 - s_4|\;\cdots\;|s_3 - s_n|}{|s_3 - s_1|\,|s_3 - s_5|\;\cdots\;|s_3 - s_m|}
\tag{7-57}
$$

Example 1. The poles and zeros of (7-19) are shown in Fig. 7-5 with the distances as marked. Since all the poles are real, we use (7-54) and write for the time function

$$
\begin{aligned}
f(t) &= \frac{(1)(1)}{(1)(2)}\,e^{-t} - \frac{(\sqrt{10}/2)(\sqrt{10}/2)}{(1)(1)}\,e^{-2t} + \frac{(\sqrt{6})(\sqrt{6})}{(2)(1)}\,e^{-3t} \\
&= \tfrac{1}{2}e^{-t} - \tfrac{5}{2}e^{-2t} + \tfrac{6}{2}e^{-3t}
\end{aligned}
\tag{7-58}
$$

The negative sign on the second term is used because there is an odd number (one) of real poles to the right.

Comparing (7-58) with the result (7-20) we see that a factor of 2 has been lost in the graphical method. This is the multiplicative constant that is always unknown when a transform is specified by its poles and zeros only. The graphical method does provide the relative magnitudes of the terms, however.

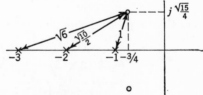

Fig. 7-5. Poles and zeros of

$$
\frac{2s^2 + 3s + 3}{(s + 1)(s + 2)(s + 3)}
$$

Suppose in this example we had known that for $s = 0$, $F(s) = \tfrac{1}{2}$ from (7-19). Now the value of $F(s)$ at $s = s'$ for

$$
F(s) = K\frac{(s - s_2)(s - s_4)\;\cdots\;(s - s_n)}{(s - s_1)(s - s_3)\;\cdots\;(s - s_m)}
\tag{7-59}
$$

is

$$
F(s') = K\frac{(s' - s_2)(s' - s_4)\;\cdots\;(s' - s_n)}{(s' - s_1)(s' - s_3)\;\cdots\;(s' - s_m)}
\tag{7-60}
$$

The terms in (7-60) can be evaluated in the same way as the coefficients —by measuring lengths and angles. In the present example we have $F(s') = \frac{1}{2}$ with $s' = 0$. From Fig. 7-6 we can determine $F(s')/K$ by multiplying distances from $s' = 0$ to zeros and dividing by the product of the distances to poles. In this case

$$\frac{F(0)}{K} = \frac{(\sqrt{6}/2)(\sqrt{6}/2)}{(1)(2)(3)} = \frac{1}{4} \tag{7-61}$$

But $F(0) = \frac{1}{2}$, so $K = 2$, and the final result is (7-58) multiplied by this factor of 2.

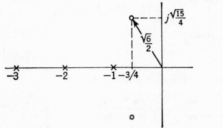

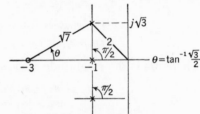

FIG. 7-6. Evaluation of $F(0)/K$.

FIG. 7-7. Poles and zeros of

$$\frac{s+3}{[(s+1)^2+3](s+1)}$$

Example 2. Let us take the poles and zeros of (7-27) as shown in Fig. 7-7. We shall also assume that $F(0)$ is known to be $\frac{3}{4}$.

The term due to the real pole at $s = -1$ is

$$f_1(t) = \frac{2}{(\sqrt{3})(\sqrt{3})} e^{-t} = \frac{2}{3} e^{-t} \tag{7-62}$$

and that due to the complex poles is, from (7-56),

$$f_2(t) = 2 \frac{\sqrt{7}}{(\sqrt{3})(2\sqrt{3})} e^{-t} \cos\left(\sqrt{3}\,t + \tan^{-1}\frac{\sqrt{3}}{2} - \frac{\pi}{2} - \frac{\pi}{2}\right)$$

$$= -\frac{\sqrt{7}}{3} e^{-t} \cos\left(\sqrt{3}\,t + \tan^{-1}\frac{\sqrt{3}}{2}\right) \tag{7-63}$$

The sum of (7-62) and (7-63) is

$$f(t) = \frac{2}{3} e^{-t}\left[1 - \frac{\sqrt{7}}{2}\cos\left(\sqrt{3}\,t + \tan^{-1}\frac{\sqrt{3}}{2}\right)\right]$$

$$= \frac{2}{3} e^{-t}\left[1 + \frac{\sqrt{7}}{2}\sin\left(\sqrt{3}\,t - \tan^{-1}\frac{2}{\sqrt{3}}\right)\right] \tag{7-64}$$

which is in agreement with (7-31). A check of $F(0)$ from Fig. 7-7 shows that it is already $\frac{3}{4}$, and no adjustment of (7-64) is needed as it was in the first example.

Second-order Poles. When the transform contains a second-order pole, the coefficient of the corresponding time term can be found graphically but the computation is more complicated. If our transform is

$$F(s) = \frac{(s - s_2)(s - s_4) \cdots (s - s_n)}{(s - s_1)^2(s - s_3) \cdots (s - s_m)} \tag{7-65}$$

the term due to the second-order pole at $s = s_1$ is [from (7-18)]

$$
\begin{aligned}
f_1(t) &= \left[\frac{d}{ds} \frac{(s - s_2)(s - s_4) \cdots (s - s_n)}{(s - s_3) \cdots (s - s_m)} \right]_{s=s_1} e^{s_1 t} \\
&\quad + \left[\frac{(s_1 - s_2)(s_1 - s_4) \cdots (s_1 - s_n)}{(s_1 - s_3) \cdots (s_1 - s_m)} \right] t e^{s_1 t} \\
&= e^{s_1 t} \left[\frac{(s_1 - s_4)(s_1 - s_6) \cdots (s_1 - s_n)}{(s_1 - s_3) \cdots (s_1 - s_m)} + \cdots \right. \\
&\quad + \frac{(s_1 - s_2)(s_1 - s_4) \cdots (s_1 - s_{n-2})}{(s_1 - s_3) \cdots (s_1 - s_m)} \\
&\quad - \frac{(s_1 - s_2)(s_1 - s_4) \cdots (s_1 - s_n)}{(s_1 - s_3)^2 \cdots (s_1 - s_m)} - \cdots \\
&\quad \left. - \frac{(s_1 - s_2)(s_1 - s_4) \cdots (s_1 - s_n)}{(s_1 - s_3) \cdots (s_1 - s_m)^2} \right] \\
&\quad + t e^{s_1 t} \left[\frac{(s_1 - s_2)(s_1 - s_4) \cdots (s_1 - s_n)}{(s_1 - s_3) \cdots (s_1 - s_m)} \right] \tag{7-66}
\end{aligned}
$$

The coefficient of $t e^{s_1 t}$ is obtained as in the case of a first-order pole. The $e^{s_1 t}$ coefficient is the sum of terms, one for each pole and zero. Let us illustrate by example.

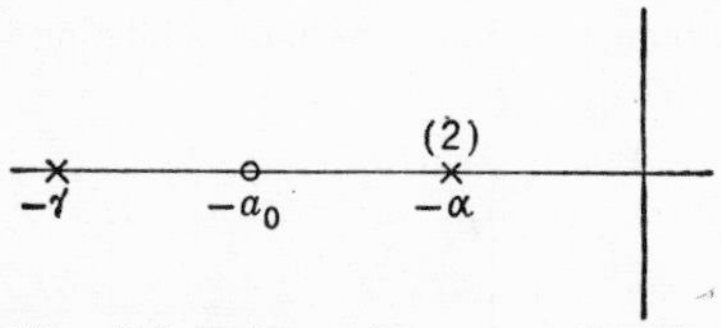

Fig. 7-8. Poles and zeros of (7-67).

Example. The poles and zeros of

$$F(s) = \frac{s + a_0}{(s + \alpha)^2(s + \gamma)} \tag{7-67}$$

are shown in Fig. 7-8. Two terms can be written at once:

$$f_1(t) = \frac{a_0 - \alpha}{\gamma - \alpha} t e^{-\alpha t} \tag{7-68}$$

$$f_2(t) = - \frac{\gamma - a_0}{(\gamma - \alpha)^2} e^{-\gamma t} \tag{7-69}$$

We note that as seen from $s = -\gamma$ the double pole at $-\alpha$ is treated as two single-order poles superimposed. The other term is, from (7-66),

$$f_3(t) = \left[\frac{1}{\gamma - \alpha} - \frac{a_0 - \alpha}{(\gamma - \alpha)^2}\right] e^{-\alpha t} \tag{7-70}$$

The inverse transform is then

$$f(t) = \frac{a_0 - \gamma}{(\gamma - \alpha)^2} e^{-\gamma t} + e^{-\alpha t}\left[\frac{a_0 - \alpha}{\gamma - \alpha} t + \frac{\gamma - a_0}{(\gamma - \alpha)^2}\right] \tag{7-71}$$

The graphical technique rapidly becomes unwieldy as higher-order poles are introduced. It may be useful as a check of a derivation by one of the other methods, however.

The graphical method is also useful in making a quick estimate of the size of a term due to a pole in $F(s)$. For example, if a zero at s_2 is very near the pole at s_1, (7-55) shows that the magnitude of the term will be small, owing to the presence of $|s_1 - s_2|$ in the numerator.

7-1. Expand in partial fractions, and then find $\mathcal{L}^{-1}$:

a. $\dfrac{1}{(s + \alpha)(s + \gamma)}$

b. $\dfrac{s + a_0}{s(s + \alpha)(s + \gamma)}$

c. $\dfrac{1}{s(s^2 + \beta^2)}$

d. $\dfrac{s^2 + a_1 s + a_0}{s(s^2 + \beta^2)}$

e. $\dfrac{1}{(s + \alpha)s^2}$

f. $\dfrac{s^2 + a_1 s + a_0}{(s + \alpha)^2 s^2}$

g. $\dfrac{1}{(s^2 + \beta^2)s^3}$

h. $\dfrac{s^3 + a_2 s^2 + a_1 s + a_0}{(s + \gamma)(s + \alpha)^2}$

7-2. Use (7-26) to work Prob. 7-1c and d above.

7-3. Tell all you can about the time functions belonging to the s-plane plots below without actually finding $f(t)$.

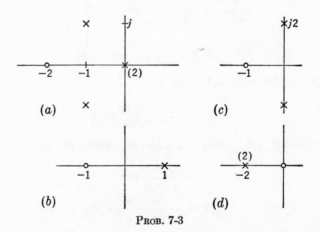

PROB. 7-3

7-4. Find $\mathcal{L}^{-1}$ in Prob. 7-1 by the residue method.

7-5. Find $\mathcal{L}^{-1}$ by residues:

a. $\dfrac{1}{s}$

b. $\dfrac{1}{s + \alpha}$

c. $\dfrac{1}{(s + \alpha)^3}$

d. $\dfrac{s + a_0}{s(s + \alpha)}$

e. $\dfrac{1}{s^n}$

f. $\dfrac{s + a_0}{(s + \alpha)^2}$

7-6. Show that a pole at $s = -\alpha - j\beta$ will produce imaginary coefficients in a rational fraction unless the fraction also has a pole at $s = -\alpha + j\beta$, the complex conjugate.

7-7. Use the graphical method to find $\mathcal{L}^{-1}$ in Prob. 7-1.

7-8. Find $\mathcal{L}^{-1}[1/s(s + \alpha)]$ as $\alpha \to 0$.

7-9. Find the inverse transform of (7-36).

CHAPTER 8

PROPERTIES OF THE LAPLACE TRANSFORM

One of the most interesting and useful parts of the study of the $\mathcal{L}$ transform is that concerned with the properties of the transform itself. In this chapter we shall discuss some of the more important properties and illustrate their uses by example.

8-1. Summary of Properties

We shall summarize the properties here, leaving the derivations and examples for subsequent sections. We adopt the notation

$$F(s) \triangleq \mathcal{L}[f(t)] \tag{8-1}$$

for what follows. The properties are presented below in transform table form, starting with (8-1):

Definitions

$f(t)$	$F(s)$	
$f_1(t)$	$F_1(s)$	(8-2)
$f(t,a)$	$F(s,a)$	

Shifting

$f(t-a)u(t-a)$		$e^{-as}F(s)$	(8-3)
$f(t+a)u(t)$	$a \geq 0$	$e^{as}\mathcal{L}[f(t)u(t-a)]$	
$e^{-at}f(t)$		$F(s+a)$	(8-4)

Change of Scale

$f(at)$	$\dfrac{1}{a}F\left(\dfrac{s}{a}\right)$	(8-5)
$a \geq 0$		

Multiplication

$f_1(t)f_2(t)$	$\dfrac{1}{2\pi j}\displaystyle\int_{c-j\infty}^{c+j\infty} F_1(w)F_2(s-w)\,dw$	(8-6)
$\displaystyle\int_0^t f_1(\tau)f_2(t-\tau)\,d\tau$	$F_1(s)F_2(s)$	(8-7)

Partial Derivatives and Integrals

$$\frac{\partial}{\partial a} f(t,a) \qquad\qquad \frac{\partial}{\partial a} F(s,a) \tag{8-8}$$

$$\int_{a_1}^{a_2} f(t,a)\, da \qquad\qquad \int_{a_1}^{a_2} F(s,a)\, da \tag{8-9}$$

Nonconstant Coefficients

$$t^n f(t) \qquad n \geq 0 \qquad\qquad (-)^n \frac{d^n}{ds^n} F(s)$$

$$\frac{1}{t} f(t) \qquad\qquad\qquad \int_s^\infty F(s)\, ds \tag{8-10}$$

Real and Imaginary Parts

$$\text{Re } [f(t)] \qquad\qquad \text{Re } [F(s)]$$
$$\text{Im } [f(t)] \qquad\qquad \text{Im } [F(s)] \tag{8-11}$$

Initial and Final Values

$$\lim_{t \to 0} f(t) = \lim_{s \to \infty} sF(s) \tag{8-12}$$

$$\lim_{t \to \infty} f(t) = \lim_{s \to 0} sF(s) \qquad \text{poles of } sF(s) \text{ in left half plane} \tag{8-13}$$

8-2. Translation in t

Let us find

$$\mathcal{L}[f(t-a)u(t-a)] = \int_0^\infty f(t-a)u(t-a)e^{-st}\, dt$$
$$= \int_a^\infty f(t-a)e^{-st}\, dt \tag{8-14}$$

If we now make a change of variable, letting $t - a = x$, (8-14) becomes

$$\mathcal{L}[f(t-a)u(t-a)] = \int_0^\infty f(x)e^{-s(x+a)}\, dx$$
$$= e^{-as} \int_0^\infty f(x)e^{-sx}\, dx$$
$$= e^{-as}F(s) \qquad a \geq 0 \tag{8-15}$$

where $F(s) \triangleq \mathcal{L}[f(t)]$. This verifies (8-3). The transform of $f(t + a)u(t)$ is found in the same way.

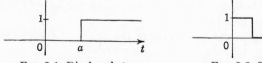

FIG. 8-1. Displaced step.   FIG. 8-2. Square wave of period T.

Example 1. A simple example of the application of (8-15) is the transform of the displaced step shown in Fig. 8-1. Since in this case $F(s) = 1/s$, the transform of the function of Fig. 8-1 is e^{-as}/s.

Example 2: Periodic Functions. We can use (8-15) to find the transform of a periodic function like the one shown in Fig. 8-2. If any $f(t)$ is periodic of period T, then by definition $f(t) = f(t + T)$. As usual, we deal with functions which are zero for negative time, so when we speak of a periodic function, we mean one which is periodic for $t \geq 0$.

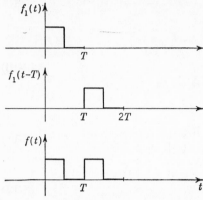

Let us define a function equal to $f(t)$ in its first period and zero elsewhere:

$$f_1(t) \triangleq \begin{cases} f(t) & 0 \leq t \leq T \\ 0 & \text{elsewhere} \end{cases} \quad (8\text{-}16)$$

Then we can write

$$f(t) = f_1(t) + f_1(t - T) + f_1(t - 2T) + \cdots \quad (8\text{-}17)$$

Figure 8-3 shows how a square wave is built up in this way. We notice that each term on the right

FIG. 8-3. Composition of square wave.

of (8-17) can be multiplied by a unit step function with the same argument as the term without changing the value of $f(t)$. We now write

$$f(t) = f_1(t) + f_1(t - T)u(t - T) + f_1(t - 2T)u(t - 2T) + \cdots \quad (8\text{-}18)$$

and transform using (8-15):

$$\begin{aligned} \mathcal{L}[f(t)] &= F_1(s) + F_1(s)e^{-Ts} + F_1(s)e^{-2Ts} + \cdots \\ &= F_1(s)[1 + e^{-Ts} + e^{-2Ts} + \cdots] \end{aligned} \quad (8\text{-}19)$$

We now recognize that the series in the brackets of (8-19) can be obtained by dividing out the expression

$$\frac{1}{1 - e^{-Ts}} = 1 + e^{-Ts} + e^{-2Ts} + \cdots \quad (8\text{-}20)$$

and finally write

$$\mathcal{L}[f(t)] = \frac{F_1(s)}{1 - e^{-Ts}} \qquad f(t) \text{ periodic with period } T \quad (8\text{-}21)$$

In using this result, it is important to remember that $F_1(s)$ is the transform of $f(t)$ in its entire first period.

We can now write down the transform of the square wave of Fig. 8-2. Here we have

$$\begin{aligned} f_1(t) &= \begin{cases} 1 & 0 < t < T/2 \\ 0 & T/2 < t \end{cases} \\ &= u(t) - u\left(t - \frac{T}{2}\right) \end{aligned} \quad (8\text{-}22)$$

Then
$$F_1(s) = \frac{1}{s} - \frac{e^{-(T/2)s}}{s} \tag{8-23}$$

Now from (8-21) we can write

$$\mathcal{L}[f(t)] = \frac{1 - e^{-(T/2)s}}{s(1 - e^{-Ts})} = \frac{1}{s[1 + e^{-(T/2)s}]} \tag{8-24}$$

It is worth noting that care must be taken in interpreting a function of the general form of (8-21) as the transform of a periodic function. The coefficient of the $1/(1 - e^{-Ts})$ must be the transform of a function which is zero for $t > T$. Unless this is true, (8-21) cannot be used for finding inverse transforms.

Example 3: Summation of Fourier Series. An interesting application of the shifting theorem was made by L. A. Pipes[1] in summing trigonometric series. For example, suppose we wish to plot the function $g(t)$:

$$g(t) = \frac{4}{\pi}\left[\frac{1}{1}\sin t + \frac{1}{3}\sin 3t + \frac{1}{5}\sin 5t + \cdots\right] \tag{8-25}$$

The transform is

$$\mathcal{L}[g(t)] = \frac{4}{\pi}\left[\frac{1}{s^2 + 1} + \frac{1}{s^2 + 9} + \frac{1}{s^2 + 25} + \cdots\right] \tag{8-26}$$

Now one series expansion for $\tanh x$ is[2]

$$\tanh x = 2x\left(\frac{2}{\pi}\right)^2\left[\frac{1}{(2x/\pi)^2 + 1} + \frac{1}{(2x/\pi)^2 + 9}\right.$$
$$\left. + \frac{1}{(2x/\pi)^2 + 25} + \cdots\right] \tag{8-27}$$

so that we can write, substituting s for $2x/\pi$,

$$\mathcal{L}[g(t)] = \frac{4}{2(\pi s/2)(2/\pi)^2\pi}\tanh\frac{\pi s}{2} = \frac{1}{s}\tanh\frac{\pi s}{2} \tag{8-28}$$

Making use of another expansion[2] for $\tanh x$:

$$\tanh x = 1 - 2e^{-2x} + 2e^{-4x} - 2e^{-6x} + \cdots \tag{8-29}$$

Eq. (8-28) becomes

$$\mathcal{L}[g(t)] = \frac{1}{s}\tanh\frac{\pi s}{2} = \frac{1}{s}[1 - 2e^{-\pi s} + 2e^{-2\pi s} - 2e^{-3\pi s} + \cdots] \tag{8-30}$$

[1] L. A. Pipes, The Summation of Fourier Series by Operational Methods, *J. Appl. Phys.*, **21**: 298–301 (April, 1950).

[2] E. P. Adams, "Smithsonian Mathematical Formulae and Tables of Elliptic Functions," p. 129, Smithsonian Institution, Washington, 1947.

Now according to (8-15) we can write $g(t)$ as a sum of translated steps:

$$g(t) = u(t) - 2u(t - \pi) + 2u(t - 2\pi) - 2u(t - 3\pi) + \cdots \quad (8\text{-}31)$$

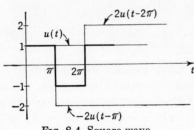

FIG. 8-4. Square wave.

The sum is shown in Fig. 8-4. It should be pointed out that the success of this method in summing series depends on finding expressions like (8-27) and (8-29).

8-3. Translation in s

The counterpart of the translation in t property of the last section can be derived by writing

$$\mathcal{L}[f(t)e^{-at}] = \int_0^\infty f(t)e^{-(s+a)t}\,dt = F(s + a) \quad (8\text{-}32)$$

where $F(s) \triangleq \mathcal{L}[f(t)]$. The property (8-32) is especially useful in deriving new transform pairs.

Example 1. Starting with the pair

$$\frac{s}{s^2 + \beta^2} \quad | \quad \cos \beta t$$

we can write immediately from (8-32)

$$\frac{s + \alpha}{(s + \alpha)^2 + \beta^2} \quad | \quad e^{-\alpha t} \cos \beta t$$

Similarly, we can start with

$$\frac{1}{s^2} \quad | \quad t$$

and get

$$\frac{1}{(s + \alpha)^2} \quad | \quad te^{-\alpha t}$$

Example 2. We can use the result (8-32) to find the transform of the product of two time functions. We shall restrict ourselves to the case where one of the functions has a transform containing first-order poles only (the general formula is derived in Sec. 8-5).

We wish to find $\mathcal{L}[f_1(t)f_2(t)]$. Now suppose the transform of $f_1(t)$ is expressed as a partial fraction

$$\mathcal{L}[f_1(t)] = F_1(s) = \frac{A}{s - s_1} + \frac{B}{s - s_2} + \cdots \quad (8\text{-}33)$$

so that

$$f_1(t) = Ae^{s_1 t} + Be^{s_2 t} + \cdots \quad (8\text{-}34)$$

Then

$$\mathcal{L}[f_1(t)f_2(t)] = \mathcal{L}[Ae^{s_1 t}f_2(t) + Be^{s_2 t}f_2(t) + \cdots] \tag{8-35}$$

and using (8-32) we can write

$$\mathcal{L}[f_1(t)f_2(t)] = AF_2(s - s_1) + BF_2(s - s_2) + \cdots \tag{8-36}$$

As an example, we use (8-36) to find $\mathcal{L}[(e^{at})^2]$. Here

$$F_1(s) = \frac{1}{s - a} \qquad A = 1 \qquad s_1 = a \tag{8-37}$$

so

$$\mathcal{L}[(e^{at})^2] = \frac{1}{(s - a) - a} = \frac{1}{s - 2a} \tag{8-38}$$

8-4. Change of Scale

Let us find

$$\mathcal{L}[f(at)] = \int_0^\infty f(at)e^{-st}\, dt$$

$$= \frac{1}{a}\int_0^\infty f(at)e^{-(s/a)at}\, d(at) = \frac{1}{a}\int_0^\infty f(x)e^{-(s/a)x}\, dx \tag{8-39}$$

The integral is the $\mathcal{L}$ transform of $f(x)$ with s replaced by s/a:

$$\mathcal{L}[f(at)] = \frac{1}{a}\,\mathcal{L}[f(t)]_{s/a} \tag{8-40}$$

If $\mathcal{L}[f(t)] \triangleq F(s)$, then from (8-40)

$$\mathcal{L}[f(at)] = \frac{1}{a}F\left(\frac{s}{a}\right) \tag{8-41}$$

Example. Starting with the transform of the unit step function:

$$u(t - 1) \qquad | \qquad \frac{e^{-s}}{s} \tag{8-42}$$

let us find $\mathcal{L}[u(t/c - 1)]$. From (8-41) we have

$$\mathcal{L}\left[u\left(\frac{t}{c} - 1\right)\right] = c\,\frac{e^{-cs}}{cs} = \frac{e^{-cs}}{s} \tag{8-43}$$

This is the same as the transform of $u(t - c)$, as we would expect, because the two functions are identical: $u(t - c) \equiv u(t/c - 1)$.

8-5. $\mathcal{L}$ Transform of a Product

In Example 2 of Sec. 8-3, we found the transform of the product of certain time functions. We shall now use the inversion formula (2-5) to find the general expression for the transform of the product of time functions:

$$\mathcal{L}^{-1}[F(s)] = \frac{1}{2\pi j}\int_{\sigma - j\infty}^{\sigma + j\infty} F(s)e^{st}\, ds \tag{8-44}$$

where $F(s) \triangleq \mathcal{L}[f(t)]$. We wish to find

$$\mathcal{L}[f_1(t)f_2(t)] = \int_0^\infty f_1(t)f_2(t)e^{-st}\, dt \tag{8-45}$$

If we write, as usual, $\mathcal{L}[f_1(t)] \triangleq F_1(s)$, etc., then (8-45) can be written, using (8-44),

$$\mathcal{L}[f_1(t)f_2(t)] = \int_0^\infty f_2(t)e^{-st}\left[\frac{1}{2\pi j}\int_{\sigma_1-j\infty}^{\sigma_1+j\infty} F_1(w)e^{wt}\, dw\right] dt \tag{8-46}$$

Interchanging order of integration (and making an adjustment of the contour) we have

$$\mathcal{L}[f_1(t)f_2(t)] = \frac{1}{2\pi j}\int_{c-j\infty}^{c+j\infty} F_1(w)\, dw \int_0^\infty f_2(t)e^{-(s-w)t}\, dt$$

$$= \frac{1}{2\pi j}\int_{c-j\infty}^{c+j\infty} F_1(w)F_2(s-w)\, dw \tag{8-47}$$

The result (8-47) is a contour integral like the one appearing in the inversion formula (8-44). A detailed discussion of this integral will be found in Appendix A.

Suppose that having (8-47) we should try to solve a nonlinear differential equation with $\mathcal{L}$ transforms. Let us try a simple one:

$$\frac{dy}{dt} + y^2 = 0 \tag{8-48}$$

The transform is, letting $\mathcal{L}[y(t)] \triangleq Y(s)$,

$$sY(s) + \frac{1}{2\pi j}\int_{c-j\infty}^{c+j\infty} Y(w)Y(s-w)\, dw = y(0) \tag{8-49}$$

Instead of reducing the differential equation to algebra, we have a more difficult equation to deal with.[1]

This example illustrates the type of difficulty encountered when we try to use $\mathcal{L}$ transforms on nonlinear equations. The complexity of (8-47) largely restricts the application of the transform method to linear equations.

8-6. The Convolution Integral

One of the most useful results in $\mathcal{L}$-transform theory is

$$\mathcal{L}^{-1}[F_1(s)F_2(s)] = \int_0^t f_1(\tau)f_2(t-\tau)\, d\tau \tag{8-50}$$

---

[1] Equation (8-48) is a form of Riccati's differential equation and has the solution

$$y(t) = \frac{y(0)}{ty(0)+1}$$

The solution of (8-49) is the transform of this function which will be a transcendental function of s. See Erdélyi, *op. cit.*, Sec. 5.12, pair (28).

where $\mathcal{L}[f_1(t)] \triangleq F_1(s)$, etc.   The integral on the right is called a *convolution integral*.

To prove (8-50), let us first note that since $f_1(t) = f_2(t) = 0$ for $t < 0$, we can write

$$\int_0^t f_1(\tau)f_2(t - \tau) \, d\tau = \int_0^\infty f_1(\tau)f_2(t - \tau) \, d\tau \qquad (8\text{-}51)$$

This follows from the fact that the argument of f_2 changes sign for $\tau > t$, making the integrand zero for those values of τ.   Now if we transform both sides of (8-50), we have

$$F_1(s)F_2(s) = \int_0^\infty e^{-st} \, dt \int_0^\infty f_1(\tau)f_2(t - \tau) \, d\tau \qquad (8\text{-}52)$$

Changing order of integration, we have

$$F_1(s)F_2(s) = \int_0^\infty f_1(\tau) \, d\tau \int_0^\infty e^{-st}f_2(t - \tau) \, dt \qquad (8\text{-}53)$$

Now since $f(t - \tau) = 0$ for $t < \tau$, we can change limits on the second integral:

$$F_1(s)F_2(s) = \int_0^\infty f_1(\tau) \, d\tau \int_\tau^\infty e^{-st}f_2(t - \tau) \, dt \qquad (8\text{-}54)$$

If we now let $t - \tau = x$, (8-54) becomes

$$F_1(s)F_2(s) = \int_0^\infty f_1(\tau) \, d\tau \int_0^\infty e^{-(x+\tau)s}f_2(x) \, dx$$
$$= \left[\int_0^\infty f_1(\tau)e^{-s\tau} \, d\tau \right] \left[\int_0^\infty f_2(x)e^{-sx} \, dx \right] \qquad (8\text{-}55)$$

Equation (8-55) is an identity, so that the original equation (8-50) is proved.

Since the order of the functions in the product $F_1(s)F_2(s)$ is unimportant, we can write

$$\mathcal{L}^{-1}[F_1(s)F_2(s)] = \int_0^t f_1(\tau)f_2(t - \tau) \, d\tau = \int_0^t f_2(\tau)f_1(t - \tau) \, d\tau \quad (8\text{-}56)$$

Example 1.   Let us use the convolution integral to find

$$\mathcal{L}^{-1}\left[\frac{1}{(s + \alpha)(s + \gamma)} \right]$$

We let

$$F_1(s) = \frac{1}{s + \alpha}$$

$$F_2(s) = \frac{1}{s + \gamma}$$

Then

$$f_1(t) = e^{-\alpha t}$$
$$f_2(t) = e^{-\gamma t} \qquad (8\text{-}57)$$

and
$$\mathcal{L}^{-1}\left[\frac{1}{(s + \alpha)(s + \gamma)}\right] = \int_0^t e^{-\alpha\tau}e^{-\gamma(t-\tau)}\,d\tau$$

$$= e^{-\gamma t}\int_0^t e^{(\gamma-\alpha)\tau}\,d\tau$$

$$= \frac{e^{-\gamma t}}{\gamma - \alpha}[e^{(\gamma-\alpha)t} - 1]$$

$$= \frac{e^{-\alpha t} - e^{-\gamma t}}{\gamma - \alpha} \tag{8-58}$$

Example 2: Graphical Interpretation of the Convolution Integral. The convolution integral (8-50) can be evaluated graphically. We write

$$\mathcal{L}^{-1}[F_1(s)F_2(s)] = \int_0^t f_1(\tau)f_2(t - \tau)\,d\tau \tag{8-59}$$

Suppose that f_1 and f_2 are as illustrated in Fig. 8-5. Now $f_2(\tau - t)$ is

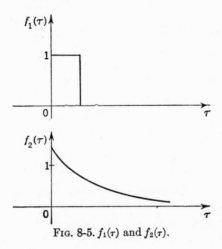

FIG. 8-5. $f_1(\tau)$ and $f_2(\tau)$.

as shown in Fig. 8-6. The function "starts" when the argument, in this case $\tau - t$, is zero. Now if we change the sign of the argument, the

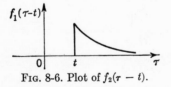

FIG. 8-6. Plot of $f_2(\tau - t)$.   FIG. 8-7. Plot of $f_2(t - \tau)$.

function is reflected about the point $\tau = t$ as shown in Fig. 8-7. The integrand of (8-59) is the product of $f_1(\tau)$ and $f_2(t - \tau)$ and will be different from zero only when the two functions overlap. In Fig. 8-8 we show the product for several values of t.   For each value of t, the integral is equal to the area of the product of f_1 and f_2. This area is indicated in our example by the shading. We can now plot these areas as shown in Fig. 8-9.

We shall have more to say about the convolution integral in the next chapter.

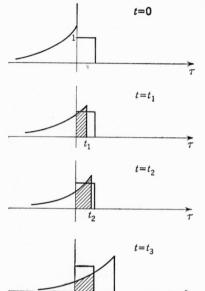

8-7. Partial Derivatives and Integrals

The basis for the application of $\mathcal{L}$ transforms to partial differential

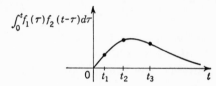

FIG. 8-8. Evaluation of convolution integral.

FIG. 8-9. Plot of convolution integral.

equations is the following property.   Suppose we wish to find

$$\mathcal{L}\left[\frac{\partial f(t,a)}{\partial a}\right] = \int_0^\infty \frac{\partial f(t,a)}{\partial a}\, e^{-st}\, dt \tag{8-60}$$

Now since a is an independent variable, we can take the partial derivative outside the integral sign and write

$$\mathcal{L}\left[\frac{\partial f(t,a)}{\partial a}\right] = \frac{\partial}{\partial a}\int_0^\infty f(t,a)e^{-st}\, dt = \frac{\partial F(s,a)}{\partial a} \tag{8-61}$$

where $F(s,a) \triangleq \mathcal{L}[f(t,a)]$.

Example 1: Partial Differential Equation. The equation which describes the motion of a vibrating string[1] is

$$\frac{\partial^2 y(x,t)}{\partial x^2} = \frac{1}{v^2}\frac{\partial^2 y(x,t)}{\partial t^2} \tag{8-62}$$

where v is the velocity of propagation of a disturbance along the string and $y(x,t)$ is the displacement of the string.   Let us transform (8-62)

---

[1] This will be derived in Sec. 11-2.

with respect to x.   Using the notation

$$\mathcal{L}_x[y(x,t)] \triangleq \int_0^\infty y(x,t)e^{-sx}\,dx \triangleq Y(s,t)$$

we have $$\mathcal{L}_x\left[\frac{\partial^2 y(x,t)}{\partial x^2}\right] = \frac{1}{v^2}\mathcal{L}_x\left[\frac{\partial^2 y(x,t)}{\partial t^2}\right] \tag{8-63}$$

Now, using (8-61), we have[1]

$$s^2 Y(s,t) - sy(0,t) - y_x(0,t) = \frac{1}{v^2}\frac{\partial^2 Y(s,t)}{\partial t^2} \tag{8-64}$$

The partial with respect to x (the derivative with t held constant) transforms like an ordinary derivative.   In (8-64) there remains only one independent variable, t, so that the equation can be written as an ordinary differential equation

$$\frac{1}{v^2}\frac{d^2 Y(s,t)}{dt^2} - s^2 Y(s,t) = -y_x(0,t) - sy(0,t) \tag{8-65}$$

We see that the $\mathcal{L}$-transform method has reduced a partial differential equation to an ordinary one—in this case with constant coefficients.   The solution of (8-65) can be found in the usual way.   We shall discuss partial differential equations further in Chap. 11.

Example 2.   We can use (8-61) to find new transform pairs.   If we start with the pair

$$e^{-at} \qquad \Big| \qquad \frac{1}{s+a}$$

and differentiate both members with respect to a, there results

$$-te^{-at} \qquad \Big| \qquad \frac{-1}{(s+a)^2}$$

Integration.   The counterpart to (8-61) is derived by starting with

$$\mathcal{L}\left[\int_{a_1}^{a_2} f(t,a)\,da\right] = \int_0^\infty e^{-st}\,dt\int_{a_1}^{a_2} f(t,a)\,da \tag{8-66}$$

Interchanging order of integration, we have

$$\mathcal{L}\left[\int_{a_1}^{a_2} f(t,a)\,da\right] = \int_{a_1}^{a_2} da\int_0^\infty e^{-st}f(t,a)\,dt = \int_{a_1}^{a_2} F(s,a)\,da \tag{8-67}$$

Example.   Let us start with the pair

$$\sin 2at \qquad \Big| \qquad \frac{2a}{s^2 + 4a^2}$$

[1] $y_x(x,t) \triangleq \dfrac{\partial y\,(x,t)}{\partial x}$.

and integrate both sides with respect to a.   We have

$$\int_0^\beta \sin 2at \, da = -\left.\frac{\cos 2at}{2t}\right|_0^\beta = \left[\frac{\sin^2 at}{t} - \frac{1}{2t}\right]_0^\beta$$
$$= \frac{\sin^2 \beta t}{t} \tag{8-68}$$

and

$$\int_0^\beta \frac{2a \, da}{s^2 + 4a^2} = \left.\frac{1}{4} \ln (s^2 + 4a^2)\right|_0^\beta = \frac{1}{4} \ln (s^2 + 4\beta^2) - \frac{1}{4} \ln s^2$$
$$= \frac{1}{4} \ln \frac{s^2 + 4\beta^2}{s^2} \tag{8-69}$$

The new pair is

$$\frac{\sin^2 \beta t}{t} \qquad \Bigg| \qquad \frac{1}{4} \ln \left[\frac{s^2 + 4\beta^2}{s^2}\right]$$

8-8. Nonconstant Coefficients

Properties which are useful when dealing with certain nonconstant-coefficient differential equations will be derived here.   Let us define, as usual, $F(s) \triangleq \mathcal{L}[f(t)]$.   We now write

$$\frac{d^n}{ds^n} F(s) = \frac{d^n}{ds^n} \int_0^\infty e^{-st} f(t) \, dt \tag{8-70}$$

Differentiating under the integral sign, we have

$$\frac{d^n}{ds^n} F(s) = (-)^n \int_0^\infty e^{-st} t^n f(t) \, dt \tag{8-71}$$

which gives us the pair

$$t^n f(t) \qquad \Bigg| \qquad (-)^n \frac{d^n}{ds^n} F(s)$$

If we now integrate $F(s)$, another relation can be derived:

$$\int_s^\infty F(s) \, ds = \int_s^\infty ds \int_0^\infty e^{-st} f(t) \, dt \tag{8-72}$$

Interchanging order of integration, we have

$$\int_s^\infty F(s) \, ds = \int_0^\infty dt \, f(t) \int_s^\infty e^{-st} \, ds = \int_0^\infty e^{-st} \frac{f(t)}{t} \, dt \tag{8-73}$$

which gives us the pair

$$\frac{f(t)}{t} \qquad \Bigg| \qquad \int_s^\infty F(s) \, ds$$

Division by higher powers of t can be handled by repeated integration of the transform.

Example 1: Laguerre's Equation.   Let us solve the linear equation

$$x \frac{d^2y}{dx^2} + (1 - x) \frac{dy}{dx} + ny = 0 \qquad y(0) = 1 \qquad (8\text{-}74)$$

This is called Laguerre's equation.   Transforming, with $\mathfrak{L}[y(x)] \triangleq Y(s)$, we have, using (8-10),

$$-\frac{d}{ds}[s^2Y(s) - sy(0) - y'(0)] + \left(1 + \frac{d}{ds}\right)[sY(s) - y(0)] + nY(s) = 0 \tag{8-75}$$

This can be written

$$(-s^2 + s)Y'(s) + (-2s + s + 1 + n)Y(s) = 0 \tag{8-76}$$

and finally we have

$$Y'(s) + \frac{s - n - 1}{s(s - 1)} Y(s) = 0 \tag{8-77}$$

which is a linear equation of the first order for $Y(s)$.   The function $Y(s)$ is the transform of the solution of Laguerre's equation (8-74).   Equation (8-77) for $Y(s)$ is solved[1] as follows:

$$\begin{aligned}
Y(s) &= K \exp\left[-\int \frac{s - n - 1}{s(s - 1)} ds\right] \\
&= K \exp\left[-\int \frac{ds}{s - 1} + (n + 1) \int \frac{ds}{s(s - 1)}\right] \\
&= K \exp\left[-\ln (s - 1) + (n + 1) \ln \frac{s - 1}{s}\right] \\
&= K \exp\left[\ln \frac{(s - 1)^n}{s^{n+1}}\right] \\
&= \frac{K(s - 1)^n}{s^{n+1}} \tag{8-78}
\end{aligned}$$

We shall postpone the evaluation of the constant K until Sec. 8.10.   We see that the $\mathfrak{L}$ transform has given us a new differential equation—in this case a simpler one—from the variable-coefficient equation.

Example 2: Bessel's Equation.   A simple form of Bessel's equation is

$$\frac{d^2y}{dt^2} + \frac{1}{t} \frac{dy}{dt} + y = 0 \qquad y(0) = 1 \qquad y'(0) = 0 \tag{8-79}$$

---

[1] See, for example, L. R. Ford, "Differential Equations," 2d ed., sec. 2-1, McGraw-Hill Book Company, Inc., New York, 1955.

Using (8-10), we transform and get

$$s^2 Y(s) - s y(0) - y'(0) + \int_s^\infty [s Y(s) - y(0)]\, ds + Y(s) = 0 \quad (8\text{-}80)$$

which becomes after rearranging terms

$$(s^2 + 1) Y(s) + \int_s^\infty [s Y(s) - 1]\, ds = s \quad (8\text{-}81)$$

If we differentiate with respect to s, this becomes

$$(s^2 + 1) Y'(s) + (2s - s) Y(s) = 0 \quad (8\text{-}82)$$

or

$$Y'(s) + \frac{s}{s^2 + 1} Y(s) = 0 \quad (8\text{-}83)$$

We have again a first-order linear equation.   The solution (which is the transform of a Bessel function) is

$$\begin{aligned}
Y(s) &= K \exp\left[-\int \frac{s\, ds}{s^2 + 1} \right] \\
&= K \exp\left[-\tfrac{1}{2} \ln (s^2 + 1) \right] \\
&= \frac{K}{\sqrt{s^2 + 1}}
\end{aligned} \quad (8\text{-}84)$$

We shall find K in Sec. 8-10.

8-9. Real and Imaginary Parts

If a time function $z(t)$ is complex, it can be written

$$z(t) = x(t) + j y(t) \quad (8\text{-}85)$$

and we call $x(t)$ the real part of $z(t)$ and $y(t)$ the imaginary part ($x(t) \triangleq$ Re $[z(t)]$, $y(t) \triangleq$ Im $[z(t)]$).   The transform of $z(t)$ is

$$\mathcal{L}[z(t)] = \int_0^\infty z(t) e^{-st}\, dt = \int_0^\infty x(t) e^{-st}\, dt + j \int_0^\infty y(t) e^{-st}\, dt \quad (8\text{-}86)$$

From this we can write

$$\begin{aligned}
\text{Re } \{\mathcal{L}[z(t)]\} &= \mathcal{L}[x(t)] = \mathcal{L}\{\text{Re } [z(t)]\} \\
\text{Im } \{\mathcal{L}[z(t)]\} &= \mathcal{L}[y(t)] = \mathcal{L}\{\text{Im } [z(t)]\}
\end{aligned} \quad (8\text{-}87)$$

When we use complex time functions in this way, s is treated as a real parameter.

Example.   Let us apply the results above to the pair

$$e^{j\beta t} \quad \Big| \quad \frac{1}{s - j\beta} = \frac{s + j\beta}{s^2 + \beta^2}$$

Taking the real part of both members gives the pair

$$\cos \beta t \qquad \Bigg| \qquad \frac{s}{s^2 + \beta^2}$$

The imaginary part gives

$$\sin \beta t \qquad \Bigg| \qquad \frac{\beta}{s^2 + \beta^2}$$

8-10. Initial and Final Values

The properties discussed so far have pertained to the effect that a change in one member of a transform pair has on the other member. The properties

$$\lim_{t \to 0} f(t) = \lim_{s \to \infty} sF(s) \tag{8-88}$$

$$\lim_{t \to \infty} f(t) = \lim_{s \to 0} sF(s) \qquad \text{poles of } sF(s) \text{ in left half plane} \tag{8-89}$$

are *equalities* and not transform pairs. To prove (8-88) we write the transform of the derivative of $f(t)$:

$$\mathcal{L}[f'(t)] = sF(s) - f(0) = \int_0^\infty f'(t)e^{-st}\, dt$$

$$= \lim_{t \to \infty} \int_0^t f'(x)e^{-sx}\, dx \tag{8-90}$$

Now if we let $s \to \infty$, then (8-90) becomes

$$\lim_{s \to \infty} sF(s) - f(0) = 0 \tag{8-91}$$

which is (8-88).

If we let $s \to 0$, then (8-90) becomes

$$\lim_{s \to 0} sF(s) - f(0) = \lim_{t \to \infty} \int_0^t f'(x)\, dx = \lim_{t \to \infty} [f(t) - f(0)] \tag{8-92}$$

which proves (8-89). The reason for the restriction on the poles of $sF(s)$ in (8-89) is to assure the existence of a limit. As discussed in Sec. 7-5, poles in the right half plane correspond to an $f(t)$ which grows without bound as $t \to \infty$. Poles on the imaginary axis mean an oscillating $f(t)$.

Example 1. Let us find the initial and final values of the current in the example of (4-17). The transform was

$$I_2(s) = \frac{-s^4 - s^3 - 2s^2 + 2s + 2}{s(s^4 + 6s^3 + 10s^2 + 8s + 4)} \tag{8-93}$$

From (8-88) and (8-89) we can write

$$i_2(0) = -1$$
$$i_2(\infty) = \tfrac{1}{2} \tag{8-94}$$

We can go further with this now and find the initial and final values of the derivative $i_2'(t)$. Writing the transform of the derivative with the help of (8-94)

$$\mathcal{L}[i_2'(t)] = sI_2(s) - i_2(0)$$
$$= \frac{-s^4 - s^3 - 2s^2 + 2s + 2}{s^4 + 6s^3 + 10s^2 + 8s + 4} + 1$$
$$= \frac{5s^3 + 8s^2 + 10s + 6}{s^4 + 6s^3 + 10s^2 + 8s + 4} \tag{8-95}$$

we have

$$i_2'(0) = 5$$
$$i_2'(\infty) = 0 \tag{8-96}$$

Example 2: Laguerre and Bessel Functions. We can now complete the solution of Laguerre's and Bessel's equations in Sec. 8-8. The solution of Laguerre's equation (8-74), which we shall call $L_n(x)$, had a transform given by (8-78):

$$Y(s) = \frac{K(s-1)^n}{s^{n+1}} \tag{8-97}$$

Now we know that $y(0) \triangleq L_n(0) = 1$, so using the initial-value theorem, we have

$$1 = L_n(0) = \lim_{s \to \infty} sY(s) = \lim_{s \to \infty} \frac{K(s-1)^n}{s^n} = K \tag{8-98}$$

and hence $K = 1$ and we have the pair

$$L_n(x) \qquad \Big| \qquad \frac{(s-1)^n}{s^{n+1}}$$

Evidently $L_0(x) = 1$, $L_1(x) = 1 - x$, etc.

The transform of the solution of Bessel's equation (8-79) is, from (8-84),

$$Y(s) = \frac{K}{\sqrt{s^2 + 1}} \tag{8-99}$$

We shall call the solution $J_0(x)$. The initial value from (8-79) is $J_0(0) = 1$. Again using the initial-value theorem we have

$$1 = J_0(0) = \lim_{s \to \infty} \frac{sK}{\sqrt{s^2 + 1}} = K \tag{8-100}$$

hence $K = 1$ and we have the pair

$$J_0(x) \qquad \Big| \qquad \frac{1}{\sqrt{s^2 + 1}}$$

8-11. Further Examples

Often several of the properties are used together. Some of the examples in this section illustrate this.

Example 1: The Sine Integral. A special function of interest in some physical problems is the sine integral:

$$\text{Si}(t) \triangleq \int_0^t \frac{\sin x}{x} \, dx \tag{8-101}$$

Starting with the pair

$$\sin x \quad \bigg| \quad \frac{1}{s^2 + 1}$$

we write a new pair from property (8-10):

$$\frac{\sin x}{x} \quad \bigg| \quad \int_s^\infty \frac{ds}{s^2 + 1}$$

Now

$$\int_s^\infty \frac{ds}{s^2 + 1} = \tan^{-1} s \bigg|_s^\infty = \frac{\pi}{2} - \tan^{-1} s = \tan^{-1} \frac{1}{s} \tag{8-102}$$

So we have the pair

$$\frac{\sin x}{x} \quad \bigg| \quad \tan^{-1} \frac{1}{s}$$

Now we use the elementary property for integration with respect to x to obtain

$$\text{Si}(t) = \int_0^t \frac{\sin x}{x} \, dx \quad \bigg| \quad \frac{1}{s} \tan^{-1} \frac{1}{s}$$

We can use the property (8-5) to generalize this:

$$\int_0^{at} \frac{\sin x}{x} \, dx \quad \bigg| \quad \frac{1}{a} \cdot \frac{a}{s} \tan^{-1} \frac{a}{s}$$

so that we have finally

$$\text{Si}(at) \quad \bigg| \quad \frac{1}{s} \tan^{-1} \frac{a}{s}$$

From the final value theorem (8-89) we have, incidentally, $\text{Si}(\infty) = \pi/2$.

Example 2: Control Area. One measure of control system performance is the area under the error-vs.-time curve when a unit step function is applied to the system.[1]

If we assume that the transform of the error $\epsilon(t)$ is a rational fraction, we can write

$$\mathcal{L}[\epsilon(t)] = \frac{A(s)}{B(s)} \tag{8-103}$$

---

[1] T. M. Stout, A Note on Control Area, *J. Appl. Phys.*, **21**:1129–1131 (November, 1950).

The area under the error curve from zero to t has the transform

$$\mathcal{L}\left[\int_0^t \epsilon(t)\, dt\right] = \frac{A(s)}{sB(s)} \tag{8-104}$$

and the total area is found from the final-value theorem (8-89):

$$\int_0^\infty \epsilon(t)\, dt = \lim_{s\to 0} s\,\frac{A(s)}{sB(s)} = \frac{A(0)}{B(0)} \tag{8-105}$$

Example 3: Moments.   The nth moment of $f(x)$, if $f(x) = 0$ for $x < 0$, is defined

$$\alpha_n \triangleq \int_0^\infty x^n f(x)\, dx \tag{8-106}$$

Starting with

$$f(x) \quad | \quad F(s)$$

we have, using the property (8-10),

$$x^n f(x) \quad | \quad (-)^n F^{(n)}(s)$$

Integration gives us

$$\int_0^t x^n f(x)\, dx \quad | \quad \frac{(-)^n}{s} F^{(n)}(s)$$

and the final-value theorem (8-89) provides α_n:

$$\alpha_n = \lim_{s\to 0} s \cdot \frac{(-)^n}{s} F^{(n)}(s) = (-)^n F^{(n)}(0) \tag{8-107}$$

Example 4: Evaluation of Integrals.   Since the $\mathcal{L}$ transform is defined as an integral, transform tables can be used as definite integral tables. The transform variable s is treated as a parameter.   For example, from the pair

$$\sin \beta t \quad | \quad \frac{\beta}{s^2 + \beta^2}$$

we can evaluate

$$\int_0^\infty e^{-\alpha t} \sin \beta t\, dt = \frac{\beta}{\alpha^2 + \beta^2} \qquad \alpha > 0 \tag{8-108}$$

where s has been replaced by α.

Similarly from the pair for the Bessel function

$$J_0(t) \quad | \quad \frac{1}{\sqrt{s^2 + 1}}$$

we have, again setting s equal to α,

$$\int_0^\infty e^{-\alpha t} J_0(t)\, dt = \frac{1}{\sqrt{\alpha^2 + 1}} \tag{8-109}$$

PROBLEMS

8-1. Sketch the functions with transforms:

    a. $\dfrac{e^{-as}}{s + \alpha}$

    b. $\dfrac{(1 - e^{-s})^2}{s}$

    c. $\dfrac{1}{s^2} - 2\dfrac{e^{-s}}{s}$

8-2. Find the $\mathcal{L}$ transform of

    a. $\delta'_{+}(t - a)$

    b. $\sin\left(t - \dfrac{\pi}{4}\right) u\left(t - \dfrac{\pi}{4}\right)$

    c. $(t - a)^2 u(t - a)$

8-3. A periodic function $(T = 1)$ is equal to e^{-t} in its first period. What is its transform?

8-4. Find the $\mathcal{L}$ transform of the periodic time functions:

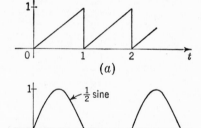

(a)

$\frac{1}{2}$ sine

(b)

Prob. 8-4

8-5. Find (using Sec. 8-3)

    a. $\mathcal{L}[e^{-(\alpha+\beta)t}]$

    b. $\mathcal{L}[te^{-\alpha t}]$

    c. $\mathcal{L}[e^{-\alpha t} \sinh t]$

8-6. Find (from Sec. 8-3)

    a. $\mathcal{L}^{-1}\left[\dfrac{1}{(s + \alpha)^3}\right]$

    b. $\mathcal{L}^{-1}\left[\dfrac{s + 2\alpha}{(s + \alpha)^2 + \beta^2}\right]$

8-7. Differentiate with respect to α to form a new pair.

    a. $\dfrac{1}{(s + \alpha)^2 + \beta^2}$

    b. $\dfrac{1}{s(s + \alpha)}$

    c. $\dfrac{1}{(s + \alpha)^2}$

8-8. Transform each partial differential equation into two ordinary differential equations (use $\mathcal{L}_t$ and $\mathcal{L}_x$).

    a. $\dfrac{\partial y}{\partial x} + \dfrac{\partial y}{\partial t} = f(t)$

    b. $\dfrac{\partial^2 y}{\partial x^2} + \dfrac{\partial^2 y}{\partial t^2} = 0$

8-9. Find the initial and final values of the functions with transforms below Check your results with the actual time functions.

    a. $\dfrac{1}{(s + \alpha)^2}$

    b. $\dfrac{1}{(s + \alpha)^2 + \beta^2}$

    c. $\dfrac{1}{(s + \alpha)(s + \gamma)}$

8-10. Transform the differential equations and solve for $y(t)$.

    a. $t\dfrac{dy}{dt} - y = 0$ *Ans.* $y = t$.

    b. $\dfrac{d^2 y}{dt^2} - t\dfrac{dy}{dt} + 2y = 0$ $y(0) = -1$ $y'(0) = 0$

                                          (Hermite's equation of degree 2)

                                         Ans. $y(t) = He_2(t) = t^2 - 1$.

8-11. Verify (8-20).

8-12. Find the approximate value of $\mathcal{L}^{-1}[1/s^2 \cdot 1/(s + \alpha)]$ by graphical convolution.

8-13. Show that $J'(0) = 0$.

8-14. Solve the integral equation

$$y(t) + e^t \int_0^t y(\tau)e^{-\tau}\, d\tau = t \qquad\qquad\qquad Ans.\ \ y = t - \frac{t^2}{2}.$$

8-15. Verify the pair

$$f(t) \sin \beta t \qquad \Big| \qquad \frac{1}{2j}\,[F(s - j\beta) - F(s + j\beta)]$$

CHAPTER 9

THE SYSTEM FUNCTION

It is characteristic of linear constant-coefficient systems that when their differential equations have been transformed, the result breaks up into three parts. One part contains terms related to the excitation, or input, to the system. A second part contains terms related to the output, or response, of the system. The third part contains terms related to the system itself. This third part is the subject of this chapter.

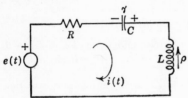

FIG. 9-1. Simple electrical system.

9-1. The System Function

Let us first study the simple electrical system shown in Fig. 9-1. Here the response will be taken as the current $i(t)$. The input causing the current flow consists of the voltage $e(t)$ and the initial conditions γ and ρ. The equation for the system is written, with the initial conditions included in the excitation,[1]

$$L \frac{di}{dt} + Ri + \frac{1}{C} \int_0^t i \, dt = e(t) + \gamma - L\rho\delta_+(t) \qquad (9\text{-}1)$$

Transforming, we have

$$\left[Ls + R + \frac{1}{Cs} \right] [I(s)] = \left[E(s) + \frac{\gamma}{s} - L\rho \right] \qquad (9\text{-}2)$$

$$\underset{\text{A}}{\qquad} \underset{\text{B}}{\qquad} \underset{\text{C}}{\qquad}$$

Equation (9-2) puts in evidence the three parts:

A: The part related to the system.

B: The part related to the response.

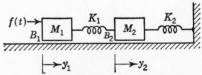

FIG. 9-2. Mechanical system.

C: The part related to the input.

The same separation occurs in higher-order systems. For example, let us consider the mechanical system shown in Fig. 9-2.

[1] See Sec. 4-3.

The equations are, again writing initial conditions as inputs,

$$M_1 \frac{d^2 y_1}{dt^2} + B_1 \frac{dy_1}{dt} + K_1 y_1 - K_1 y_2$$
$$= f(t) + [M_1 \dot{y}_1(0) + B_1 y_1(0)]\delta_+(t) + M_1 y_1(0)\delta'_+(t)$$
$$-K_1 y_1 + M_2 \frac{d^2 y_2}{dt^2} + B_2 \frac{dy_2}{dt} + (K_1 + K_2) y_2$$
$$= [M_2 \dot{y}_2(0) + B_2 y_2(0)]\delta_+(t) + M_2 y_2(0)\delta'_+(t) \tag{9-3}$$

The transformed equations can be written

$$\underbrace{\begin{bmatrix} M_1 s^2 + B_1 s + K_1 & -K_1 \\ -K_1 & M_2 s^2 + B_2 s + (K_1 + K_2) \end{bmatrix}}_{A} \underbrace{\begin{bmatrix} Y_1(s) \\ Y_2(s) \end{bmatrix}}_{B}$$

$$= \underbrace{\begin{bmatrix} F(s) + M_1 \dot{y}_1(0) + (M_1 s + B_1) y_1(0) \\ M_2 \dot{y}_2(0) + (M_2 s + B_2) y_2(0) \end{bmatrix}}_{C} \tag{9-4}$$

Again the transformed equations break into three parts. In this case there are two possible outputs, y_1 and y_2, and several inputs.

Let us now define a function which is characteristic of the system, regardless of the form of the input:

The system function[1] $W(s)$ is defined by

$$W(s) \triangleq \frac{\mathcal{L} \, [\text{output}]}{\mathcal{L} \, [\text{input}]} \tag{9-5}$$

where a single input is applied, all other inputs being set to zero.

In the case of the electrical system of Fig. 9-1, the transfer function is, from (9-2),

$$W(s) = \frac{1}{Ls + R + (1/Cs)} \tag{9-6}$$

The system of Fig. 9-2 has two coordinates. The possible system functions are:

1. $W_{11}(s)$ relating y_1 to forces applied to M_1 (including initial conditions).

2. $W_{22}(s)$ relating y_2 to forces applied to M_2 (including initial conditions).

3. $W_{12}(s)$ relating y_1 to forces applied to M_2 (including initial conditions).

4. $W_{21}(s)$ relating y_2 to forces applied to M_1 (including initial conditions).

---

[1] Another term sometimes used is *transfer function.*

From (9-4) we have

$$Y_1(s) = \frac{\begin{vmatrix} F_1(s) & -K_1 \\ F_2(s) & M_2s^2 + B_2s + (K_1 + K_2) \end{vmatrix}}{\begin{vmatrix} M_1s^2 + B_1s + K_1 & -K_1 \\ -K_1 & M_2s^2 + B_2s + (K_1 + K_2) \end{vmatrix}}$$

$$Y_2(s) = \frac{\begin{vmatrix} M_1s^2 + B_1s + K_1 & F_1(s) \\ -K_1 & F_2(s) \end{vmatrix}}{\begin{vmatrix} M_1s^2 + B_1s + K_1 & -K_1 \\ -K_1 & M_2s^2 + B_2s + (K_1 + K_2) \end{vmatrix}} \qquad (9\text{-}7)$$

where

$$\begin{aligned} F_1(s) &\triangleq F(s) + M_1\dot{y}_1(0) + (M_1s + B_1)y_1(0) \\ F_2(s) &\triangleq M_2\dot{y}_2(0) + (M_2s + B_2)y_2(0) \end{aligned} \qquad (9\text{-}8)$$

We now write

$$W_{11}(s) = \left[\frac{Y_1(s)}{F_1(s)}\right]_{F_2(s)=0} = \frac{M_2s^2 + B_2s + K_1 + K_2}{(M_1s^2 + B_1s + K_1)(M_2s^2 + B_2s + K_1 + K_2) - K_1{}^2} \qquad (9\text{-}9)$$

Similarly

$$W_{22}(s) = \left[\frac{Y_2(s)}{F_2(s)}\right]_{F_1(s)=0} = \frac{M_1s^2 + B_1s + K_1}{(M_1s^2 + B_1s + K_1)(M_2s^2 + B_2s + K_1 + K_2) - K_1{}^2} \qquad (9\text{-}10)$$

and

$$W_{12}(s) = \left[\frac{Y_1(s)}{F_2(s)}\right]_{F_1(s)=0} = \frac{K_1}{(M_1s^2 + B_1s + K_1)(M_2s^2 + B_2s + K_1 + K_2) - K_1{}^2} \qquad (9\text{-}11)$$

and

$$W_{21}(s) = \left[\frac{Y_2(s)}{F_1(s)}\right]_{F_2(s)=0} = W_{12}(s) \qquad (9\text{-}12)$$

Now from the definition of the system function (9-5) we can write for the transform of the current in Fig. 9-1

$$I(s) = E_1(s)W(s) \qquad (9\text{-}13)$$

where

$$E_1(s) \triangleq E(s) + \frac{\gamma}{s} - L\rho \qquad (9\text{-}14)$$

and $W(s)$ is given by (9-6). Equation (9-13) is a general expression valid for any input applied at $t = 0$, and the inverse transform will give the response of the system to this general input for $t > 0$.

In the mechanical system of Fig. 9-2 we have, using the definition of the system function and the superposition property (Sec. 1-3),

$$\begin{aligned} Y_1(s) &= F_1(s)W_{11}(s) + F_2(s)W_{12}(s) \\ Y_2(s) &= F_1(s)W_{21}(s) + F_2(s)W_{22}(s) \end{aligned} \qquad (9\text{-}15)$$

Equations (9-15) give the transforms of the displacements y_1 and y_2 in response to general inputs applied at $t = 0$.

9-2. Block Diagrams

The equation relating input, output, and system function

$$F_0(s) = W(s)F_i(s) \tag{9-16}$$

where
$$F_0(s) \triangleq \mathcal{L} \text{ [output]}$$
$$F_i(s) \triangleq \mathcal{L} \text{ [input]} \tag{9-17}$$

is often represented by a block diagram as shown in Fig. 9-3.

FIG. 9-3. Block diagram of a system. 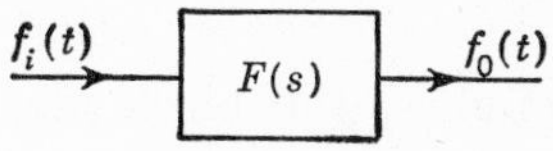   FIG. 9-4. Block diagram with time functions as input and output.

We always include initial conditions of the input coordinate with the input to that coordinate and set all others to zero when writing equations in system function form. Therefore, the transform of the derivative of $f(t)$ is simply $sF(s)$. This leads us sometimes to label the block diagram with time functions as in Fig. 9-4, keeping in mind the interpretation of the s in the system function as the d/dt operator.

In the example of Fig. 9-1, we can represent the system as shown in Fig. 9-5 [using (9-6) and (9-13)]. Alternately, we might label the block diagram as in Fig. 9-6.

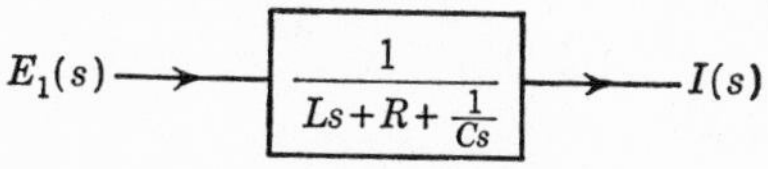

$$E_1(s) \longrightarrow \boxed{\frac{1}{Ls+R+\frac{1}{Cs}}} \longrightarrow I(s)$$

FIG. 9-5. System of Fig. 9-1.

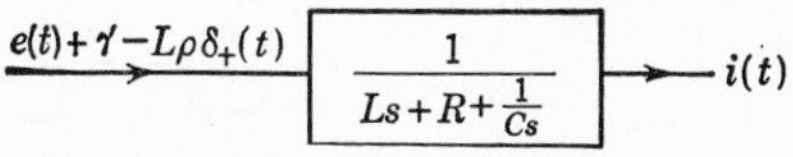

$$\frac{e(t)+\gamma-L\rho\delta_+(t)}{} \longrightarrow \boxed{\frac{1}{Ls+R+\frac{1}{Cs}}} \longrightarrow i(t)$$

FIG. 9-6. Alternate representation of system of Fig. 9-1.

The original equation can be retrieved from the block diagram. Interpreting s as d/dt and $1/s$ as the integral, we write from Fig. 9-6

$$i(t) = \left[\frac{1}{Ls + R + (1/Cs)} \right] [e(t) + \gamma - L\rho\delta_+(t)] \tag{9-18}$$

Multiplying out, we have

$$L\frac{di}{dt} + Ri + \frac{1}{C}\int_0^t i\,dt = e(t) + \gamma - L\rho\delta_+(t) \tag{9-19}$$

which is the original equation (9-1).

In this simple one-coordinate example it was possible to include initial conditions with the system input. In general, however, a single system input cannot furnish arbitrary initial conditions within the system. *We therefore assume when using system functions that all initial conditions within the system are zero.*

The system of Fig. 9-2 has an interesting block diagram derived from (9-15) and shown in Fig. 9-7. The symbol for addition is the circle in

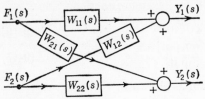

Fig. 9-7. It is important to show the direction of flow by arrows. Notice that a second input $F_2(s)$ is required for the initial conditions on M_2.

Sometimes the output of one system is used for the input of another. If the two systems do

FIG. 9-7. Block diagram for Fig. 9-2.

not interact, the block diagrams can be connected in tandem. The combined system function is the product of individual functions as shown in Fig. 9-8.

As an example, let us consider the system shown in Fig. 9-9. The isolating amplifier has unity gain, very high input impedance, and very low output impedance. This means that the network on the left can be

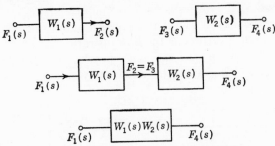

FIG. 9-8. Block diagram for combined systems.

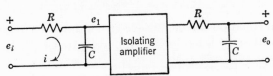

FIG. 9-9. Combined system with isolation.

considered as a single loop, since negligible current flows into the amplifier. Also the input voltage to the network on the right will be the output voltage of the left-hand network. The transfer function for both networks is found by writing

$$Ri + \frac{1}{C} \int_0^t i \, dt = e_i$$

$$e_1 = \frac{1}{C} \int_0^t i \, dt \tag{9-20}$$

Transforming, we have

$$\left[R + \frac{1}{Cs} \right] I(s) = E_i(s)$$

$$E_1(s) = \frac{1}{Cs} I(s)$$

(9-21)

Combining, we have finally

$$W(s) = \frac{E_1(s)}{E_i(s)} = \frac{1/Cs}{R + (1/Cs)} = \frac{1}{RCs + 1}$$

(9-22)

The block diagram for Fig. 9-9 is shown in Fig. 9-10.   If we had omitted

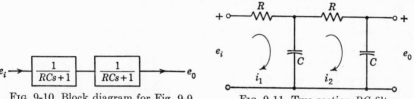

FIG. 9-10. Block diagram for Fig. 9-9.   FIG. 9-11. Two-section RC filter.

the isolation amplifier, the network would have been as shown in Fig. 9-11.   In this case the equations are

$$R i_1 + \frac{1}{C} \int_0^t i_1 \, dt - \frac{1}{C} \int_0^t i_2 \, dt = e_i$$

$$- \frac{1}{C} \int_0^t i_1 \, dt + R i_2 + \frac{2}{C} \int_0^t i_2 \, dt = 0$$

(9-23)

$$e_0 = \frac{1}{C} \int_0^t i_2 \, dt$$

Transforming, we have

$$\begin{bmatrix} R + \dfrac{1}{Cs} & - \dfrac{1}{Cs} \\[2ex] - \dfrac{1}{Cs} & R + \dfrac{2}{Cs} \end{bmatrix} \begin{bmatrix} I_1(s) \\[2ex] I_2(s) \end{bmatrix} = \begin{bmatrix} E_i(s) \\[2ex] 0 \end{bmatrix}$$

(9-24)

$$E_0(s) = \frac{1}{Cs} I_2(s)$$

from which we have

$$I_2(s) = \frac{\begin{vmatrix} R + \dfrac{1}{Cs} & E_i(s) \\[2ex] - \dfrac{1}{Cs} & 0 \end{vmatrix}}{\begin{vmatrix} R + \dfrac{1}{Cs} & - \dfrac{1}{Cs} \\[2ex] - \dfrac{1}{Cs} & R + \dfrac{2}{Cs} \end{vmatrix}} = \frac{\dfrac{E_i}{Cs}}{\left(R + \dfrac{1}{Cs} \right) \left(R + \dfrac{2}{Cs} \right) - \dfrac{1}{C^2 s^2}}$$

(9-25)

so that finally

$$W(s) = \frac{E_0(s)}{E_i(s)} = \frac{I_2(s)/Cs}{E_i(s)} = \frac{1}{C^2 s^2 [R + (1/Cs)][R + (2/Cs)] - 1} \quad (9\text{-}26)$$

Evidently interaction between the RC sections has changed the system function from the one obtained in Fig. 9-9 where the sections were isolated.

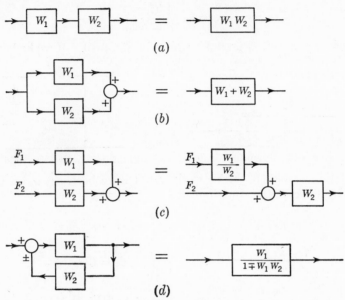

FIG. 9-12. Manipulation of block diagrams.

A number of useful manipulations[1] possible with block diagrams are shown in Fig. 9-12. These can be verified from the definition of the system function (9-16). The feedback system (Fig. 9-12d) will be discussed in detail in the next chapter.

9-3. Impulse Response

We have seen that for a linear, constant-coefficient system it is possible to define a system function $W(s)$ from which the response of the system to a general input can be obtained. Let us suppose now that we apply a unit impulse function $\delta_+(t)$ to the system shown in Fig. 9-13. We shall call the

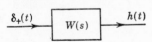

FIG. 9-13. Unit impulse applied to a system.

impulse response of the system $h(t)$. Now from (9-16) we can write

$$\mathcal{L}[h(t)] = W(s) \cdot \mathcal{L}[\delta_+(t)] \quad (9\text{-}27)$$

[1] For a detailed discussion see J. G. Truxal, "Automatic Feedback Control System Synthesis," chap. 2, McGraw-Hill Book Company, Inc., New York, 1955.

But the transform of the impulse is unity, so we have

$$W(s) = \mathcal{L}[h(t)] \tag{9-28}$$

This is a very useful result because we can use it to find the response of the system to any input applied at $t = 0$.

Starting with the defined relation for $W(s)$

$$F_0(s) = W(s)F_i(s) \tag{9-29}$$

we can write

$$f_0(t) = \mathcal{L}^{-1}[W(s)F_i(s)] \tag{9-30}$$

But the inverse transform of a product is the convolution integral (Sec. 8-6), so we can write (9-30), using (9-28), in either of the forms

$$f_0(t) = \int_0^t h(t - \tau)f_i(\tau)\,d\tau = \int_0^t h(\tau)f_i(t - \tau)\,d\tau \tag{9-31}$$

Because the impulse response in the integrand gives varying importance, or weight, to the input f_i, the impulse response is sometimes called the *weighting function.*[1]

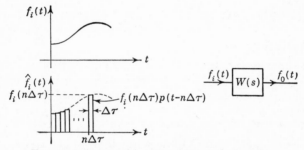

FIG. 9-14. A general input and its approximation.

The result (9-31) can also be obtained from a physical argument. Suppose a general $f_i(t)$ is applied to a system as shown in Fig. 9-14. Let us approximate $f_i(t)$ by a function $\hat{f}_i(t)$ made up of narrow pulses of width $\Delta\tau$ as shown in the figure. Because of the superposition property of linear systems (Sec. 1-3), the response $\hat{f}_0(t)$ to the approximating function is the sum of the responses to the individual pulses making it up. If we call the pulse (of unit height and width $\Delta\tau$ beginning at $t = 0$) $p(t)$, then the approximate input is given by

$$\hat{f}_i(t) = \sum_{n=0}^{\infty} p(t - n\,\Delta\tau)f_i(n\,\Delta\tau) \tag{9-32}$$

[1] The impulse response is also identical with the *Green's function.* See B. Friedman, "Principles and Techniques of Applied Mathematics," chap. 3, John Wiley & Sons, Inc., New York, 1956.

Then if the response of the system to $p(t)$ is $h_1(t)$, the output can be written as the sum

$$\hat{f}_0(t) = \sum_{n=0}^{\infty} h_1(t - n\,\Delta\tau)f_i(n\,\Delta\tau) \tag{9-33}$$

Now as $\Delta\tau$, the width of the pulse, is made smaller, the response approaches the response to an impulse of area $\Delta\tau$ [since $p(t)$ is of unit height]

$$h_1(t - n\,\Delta\tau) \rightarrow h(t - n\,\Delta\tau)\,\Delta\tau \tag{9-34}$$

If we hold $n\,\Delta\tau = \tau$ as $\Delta\tau \rightarrow 0$, the sum in (9-33) becomes an integral and we have

$$f_0(t) = \lim_{\substack{\Delta\tau \rightarrow 0 \\ n\,\Delta\tau = \tau}} \hat{f}_0(t) = \int_0^{\infty} h(t - \tau)f_i(\tau)\,d\tau \tag{9-35}$$

But since $h(t - \tau) = 0$ for $\tau > t$, (9-35) becomes

$$f_0(t) = \int_0^{t} h(t - \tau)f_i(\tau)\,d\tau \tag{9-36}$$

which agrees with (9-31).

9-4. Frequency Response

We shall show in this section how the system function is related to the steady-state response of the system to a sinusoidal input.

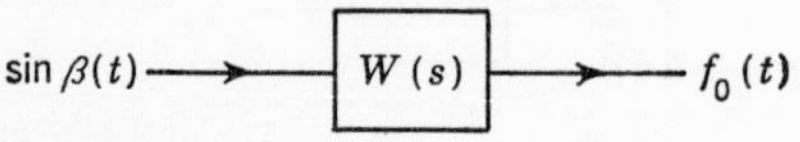

$\sin \beta(t) \longrightarrow \boxed{W(s)} \longrightarrow f_0(t)$

Fig. 9-15. System with sinusoidal input.

Let us suppose that the system in Fig. 9-15 is subjected to a unit amplitude sinusoidal input. If we write $W(s) = A(s)/B(s)$, the transform of the output is

$$F_0(s) = W(s)\frac{\beta}{s^2 + \beta^2} = \frac{\beta A(s)}{B(s)(s^2 + \beta^2)} \tag{9-37}$$

We can make a partial-fraction expansion of (9-37) using the results we obtained in Sec. 7-3. From (7-26) we can write

$$F_0(s) = W(s)\frac{\beta}{s^2 + \beta^2} = \beta \operatorname{Re}\left[\frac{W(j\beta)}{j\beta(s - j\beta)}\right] + \frac{C(s)}{B(s)} \tag{9-38}$$

where the function $C(s)/B(s)$ has the same poles as $W(s)$. Taking the inverse transform of (9-38), we have

$$f_0(t) = \mathcal{L}^{-1}\left\{\operatorname{Re}\left[\frac{W(j\beta)}{j(s - j\beta)}\right]\right\} + \mathcal{L}^{-1}\left[\frac{C(s)}{B(s)}\right] \tag{9-39}$$

We now assume that $W(s)$ has poles only in the left half plane, so that the transients due to application of the input will die out with time. (Such a system is *stable*. See Sec. 7-5.) This means that the second term on the right in (9-39) will go to zero as $t \to \infty$, leaving the steady-state part of the output:

$$f_{0_{ss}}(t) = \mathcal{L}^{-1} \left\{ \operatorname{Re} \left[\frac{W(j\beta)}{j(s - j\beta)} \right] \right\} \tag{9-40}$$

Using the property of Sec. 8-9 to interchange $\mathcal{L}^{-1}$ and Re operators, we have

$$f_{0_{ss}}(t) = \operatorname{Re} \left\{ \mathcal{L}^{-1} \left[\frac{W(j\beta)}{j(s - j\beta)} \right] \right\} = \operatorname{Re} \left[\frac{W(j\beta)}{j} e^{j\beta t} \right] \tag{9-41}$$

Let us now write $W(j\beta)$ in polar form:

$$W(j\beta) = |W(j\beta)|e^{j\psi} \tag{9-42}$$

so that we have finally

$$f_{0_{ss}} = |W(j\beta)| \operatorname{Re} \left[\frac{e^{j(\beta t + \psi)}}{j} \right] = |W(j\beta)| \sin (\beta t + \psi) \tag{9-43}$$

The angle ψ of $W(j\beta)$ is the *phase angle* of the response; $|W(j\beta)|$ is the *magnitude*.

The relation between system function and frequency response (9-43) makes it easy to find steady-state response for a sinusoidal input. We shall discuss this means of describing system behavior further in the next section.

Impulse and Frequency Response. From the relation between system function and impulse response (9-28) we can get a relation between frequency response and impulse response. Starting with

$$W(s) = \mathcal{L}[h(t)] = \int_0^\infty e^{-st} h(t)\, dt \tag{9-44}$$

we can write

$$W(j\beta) = \int_0^\infty e^{-j\beta t} h(t)\, dt = \int_0^\infty \cos \beta t\, h(t)\, dt - j \int_0^\infty \sin \beta t\, h(t)\, dt \tag{9-45}$$

From (9-45) $W(j\beta)$ can be obtained by numerical integration from an experimentally determined $h(t)$. The relation (9-45) will be valid when $W(s)$ has poles only in the left half plane, assuring an $h(t)$ which dies out exponentially with time.

We can also invert (9-44) to obtain $h(t)$ from frequency-response data.[1] We write the inversion formula (2-5) for (9-44):

$$h(t) = \frac{1}{2\pi j} \int_{-j\infty}^{j\infty} W(s) e^{st}\, ds \tag{9-46}$$

---

[1] E. A. Guillemin, "Communication Networks," vol. 2, Chap. 11, Sec. 8, John Wiley & Sons, Inc., New York, 1935.

where we have set $\sigma = 0$ because of the stable behavior of $h(t)$. Now, letting $s = j\beta$, (9-46) becomes

$$h(t) = \frac{1}{2\pi} \int_{-\infty}^{\infty} W(j\beta) e^{j\beta t} \, d\beta \qquad (9\text{-}47)$$

If we now write $W(j\beta)$ in polar form, we have[1]

$$
\begin{aligned}
h(t) &= \frac{1}{2\pi} \int_{-\infty}^{\infty} |W(j\beta)| e^{j\psi} e^{j\beta t} \, d\beta \\
&= \frac{1}{2\pi} \int_{-\infty}^{\infty} |W(j\beta)| (\cos \psi + j \sin \psi)(\cos \beta t + j \sin \beta t) \, d\beta \\
&= \frac{1}{2\pi} \int_{-\infty}^{\infty} |W(j\beta)| (\cos \psi \cos \beta t - \sin \psi \sin \beta t) \, d\beta \\
&\quad + \frac{j}{2\pi} \int_{-\infty}^{\infty} |W(j\beta)| (\cos \psi \sin \beta t + \sin \psi \cos \beta t) \, d\beta \quad (9\text{-}48)
\end{aligned}
$$

Now $h(t)$ is real, so that the second integral in (9-48), which has the coefficient j, must be zero. We now have

$$h(t) = \frac{1}{2\pi} \int_{-\infty}^{\infty} |W(j\beta)| \cos \psi \cos \beta t \, d\beta - \frac{1}{2\pi} \int_{-\infty}^{\infty} |W(j\beta)| \sin \psi \sin \beta t \, d\beta \quad (9\text{-}49)$$

Now $h(t)$, the impulse response, is zero for negative time, so we can write

$$h(-t) = \frac{1}{2\pi} \int_{-\infty}^{\infty} |W(j\beta)| \cos \psi \cos \beta t \, d\beta + \frac{1}{2\pi} \int_{-\infty}^{\infty} |W(j\beta)| \sin \psi \sin \beta t \, d\beta = 0 \quad (9\text{-}50)$$

The two integrals in (9-49) are evidently equal except for sign, so we can write (9-49)

$$h(t) = \frac{1}{\pi} \int_{-\infty}^{\infty} |W(j\beta)| \cos \psi \cos \beta t \, d\beta \qquad (9\text{-}51)$$

But $|W(j\beta)|$ is an even function of β

$$|W(j\beta)| = [W(j\beta) W(-j\beta)]^{1/2} = |W(-j\beta)| \qquad (9\text{-}52)$$

as are $\cos \beta t$ and $\cos \psi$, so we have finally

$$h(t) = \frac{2}{\pi} \int_{0}^{\infty} |W(j\beta)| \cos \psi \cos \beta t \, d\beta \qquad (9\text{-}53)$$

[1] Note that ψ is a function of β.

Measured frequency-response data can be used in (9-53) to determine the impulse response. Graphical techniques have been developed by Floyd[1] for the evaluation of both (9-45) and (9-53).

9-5. Plotting Frequency Response

In this section we shall discuss a means for plotting frequency response. By *frequency response* we mean the steady-state response of a system to a unit amplitude sinusoidal input. In the preceding section we saw that if we write the system function

$$W(j\omega) = |W(j\omega)|e^{j\psi} \tag{9-54}$$

then $|W(j\omega)|$ is the amplitude of the response at frequency ω radians/sec and ψ is the phase angle. The frequency response, then, consists of two parts. We shall discuss the amplitude part first:

Amplitude Response. We can interpret $|W(j\omega)|$ as the ratio of output amplitude to input amplitude. In the result (9-43) we have

$$\frac{\text{Output amplitude}}{\text{Input amplitude}} = \frac{|W(j\omega)|}{1} = |W(j\omega)| \tag{9-55}$$

Since the system is linear, the relation holds for any input amplitude. The superposition property assures that if the input is multiplied by a constant factor, the output will be multiplied by the same factor.

Now let us write down a general system function in a form which will be convenient for plotting:[2]

$$W(s) = K \frac{(\tau_1 s + 1)\left[\dfrac{s^2}{\omega_3{}^2} + 2\dfrac{\zeta_3}{\omega_3}s + 1\right]\cdots}{s^n(\tau_2 s + 1)\left[\dfrac{s^2}{\omega_4{}^2} + 2\dfrac{\zeta_4}{\omega_4}s + 1\right]\cdots} \tag{9-56}$$

We have factored numerator and denominator into terms with real roots and quadratic terms with complex roots.

We wish to find the amplitude $|W(j\omega)|$. We write

$$|W(j\omega)| = \frac{K(\tau_1{}^2\omega^2 + 1)^{1/2}\left[\left(1 - \dfrac{\omega^2}{\omega_3{}^2}\right)^2 + 4\dfrac{\zeta_3{}^2}{\omega_3{}^2}\omega^2\right]^{1/2}\cdots}{\omega^n(\tau_2{}^2\omega^2 + 1)^{1/2}\left[\left(1 - \dfrac{\omega^2}{\omega_4{}^2}\right)^2 + \dfrac{4\zeta_4{}^2}{\omega_4{}^2}\omega^2\right]^{1/2}\cdots} \tag{9-57}$$

---

[1] Brown and Campbell, "Principles of Servomechanisms," chap. 11, John Wiley & Sons, Inc., New York, 1948.

[2] Common terminology for transfer functions written in this form: τ = time constant; ω = undamped natural frequency; ζ = damping ratio.

If we now take the $\log_{10}$ of both sides and multiply by 20, we have

$$20 \log_{10} |W(j\omega)| = 20 \log_{10} \frac{K}{\omega^n} + 10 \log_{10} (1 + \omega^2 \tau_1^2)$$

$$+ 10 \log_{10} \left[\left(1 - \frac{\omega^2}{\omega_3^2} \right)^2 + \frac{4\zeta_3^2}{\omega_3^2} \omega^2 \right] + \cdots$$

$$- 10 \log_{10} (1 + \omega^2 \tau_2^2) - 10 \log_{10} \left[\left(1 - \frac{\omega^2}{\omega_4^2} \right)^2 + \frac{4\zeta_4}{\omega_4^2} \omega^2 \right] - \cdots \quad (9\text{-}58)$$

The left side is simply $|W(j\omega)|$ measured in decibels.[1] The right side is the sum of terms of three types. Since we are to add these terms, we can plot them separately and perform the addition graphically. We shall now discuss each of the three types of terms.

1. We wish to plot

$$A(\omega) = 20 \log_{10} \frac{K}{\omega^n} = 20 \log_{10} K - 20n \log_{10} \omega \quad (9\text{-}59)$$

If we plot $A(\omega)$ vs. $\log_{10} \omega$—which is conveniently done on semilog paper—the curve will be a straight line. The slope will be[2]

$$\frac{dA(\omega)}{d(\log_{10} \omega)} = -20n \quad \text{db/decade} \quad (9\text{-}60)$$

where in choosing units we have used the fact that a unit change in $\log_{10} \omega$ occurs each time ω is changed by a factor of 10. The line passes through zero when $\omega = K^{1/n}$. The plot is shown in Fig. 9-16.

2. Next, let us plot

$$B(\omega) = 10 \log_{10} (1 + \omega^2 \tau^2) \quad (6\text{-}61)$$

When $\omega \ll 1/\tau$, we have

$$B(\omega) \cong 10 \log_{10} (1) = 0 \quad (9\text{-}62)$$

and when $\omega \gg 1/\tau$,

$$B(\omega) \cong 10 \log_{10} (\omega^2 \tau^2) = 20 \log_{10} \omega\tau \quad (9\text{-}63)$$

---

[1] The *decibel* was originally defined as 10 times $\log_{10}$ of a power ratio. This led to its use for voltage ratios

$$10 \log_{10} \frac{V_1^2/R}{V_2^2/R} = 20 \log_{10} \frac{V_1}{V_2}$$

It is now commonly used for amplitude ratios of all kinds—voltage, angle, displacement, etc.

[2] The slope is sometimes measured in decibels per octave—an octave being a change by a factor of 2. The number x of octaves in a decade is found by writing $2^x = 10$, giving $x = 1/\log_{10} 2$. A slope of 20 db/decade corresponds to

$$\frac{20}{\log_{10} 2} = 6.02 \approx 6 \text{ db/octave}$$

In the case of small ω, $B(\omega)$ is zero. For large ω, it is a straight line of slope 20 db/decade, passing through zero when $\omega = 1/\tau$ [compare (9-63) with (9-59), setting $n = -1$, $K = \tau$]. The straight-line approximation is plotted in Fig. 9-17.

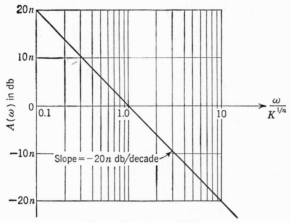

Fig. 9-16. Plot of $A(\omega)$.

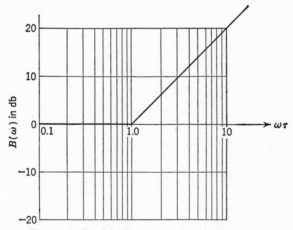

Fig. 9-17. Straight-line approximation for $B(\omega)$.

An actual plot obtained from (9-61) is shown in Fig. 9-18. Comparison with the straight-line approximation shows that the greatest error is only 3 db.

3. The plot of the quadratic term

$$C(\omega) = 10 \log_{10}\left[\left(1 - \frac{\omega^2}{\omega_0^2}\right)^2 + \frac{4\zeta^2}{\omega_0^2}\omega^2\right] \tag{9-64}$$

can be approximated by straight lines, but values of ζ much different from unity cause large errors in the approximation as can be seen from

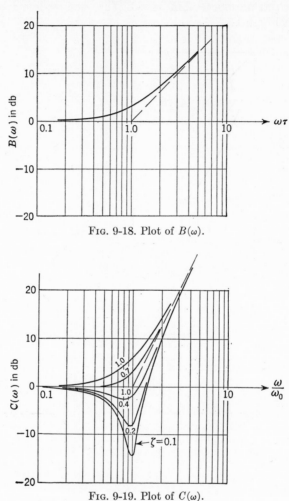

FIG. 9-18. Plot of $B(\omega)$.

FIG. 9-19. Plot of $C(\omega)$.

Fig. 9-19.   In Fig. 9-19 the straight-line approximation is shown as a dashed line.   The approximation for large ω is found by writing

$$C(\omega) \cong 10 \log_{10} \frac{\omega^4}{\omega_0{}^4} = 40 \log_{10} \frac{\omega}{\omega_0} \qquad \omega \gg \omega_0 \qquad (9\text{-}65)$$

which is a straight line of slope 40 db/decade, passing through zero when $\omega = \omega_0$.   For small values of ω we have

$$C(\omega) \cong \log_{10} (1) = 0 \qquad \omega \ll \omega_0 \qquad (9\text{-}66)$$

The curves in Fig. 9-20 give the peak value of $|C(\omega)|$ and the frequency at which it occurs. Figure 9-21 gives the value of $C(\omega)$ when $\omega = \omega_0$. These can be used in conjunction with the straight-line approximation to sketch the amplitude response.

Amplitude-response plots like the ones we have been making are called *log-db plots* or *Bode plots*.[1] The advantages of using the graphical technique for obtaining amplitude response are:

a. The straight-line approximations are often good enough.

b. More accurate plots can be made from standard curves (Figs. 9-18, 9-19, 9-20, and 9-21).

c. Contributions from various terms in a system function are added graphically.

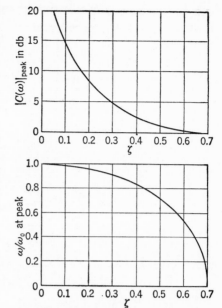

Fig. 9-20. Size and location of peak of

$$C(\omega) = 10 \log_{10}\left[\left(1 - \frac{\omega^2}{\omega_0^2}\right)^2 + 4\zeta^2 \frac{\omega^2}{\omega_0^2}\right]$$

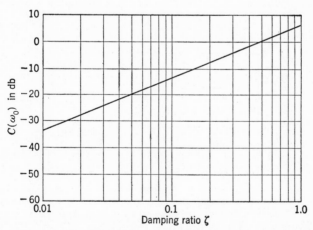

Fig. 9-21. Value of $C(\omega)$ for $\omega = \omega_0$.

d. The plot for $|F(j\omega)|$ in decibels is the negative of the plot for $1/|F(j\omega)|$ in decibels.

Figure 9-22 is a summary of amplitude responses.

[1] H. W. Bode, "Network Analysis and Feedback Amplifier Design," D. Van Nostrand Company, Inc., Princeton, N.J., 1945.

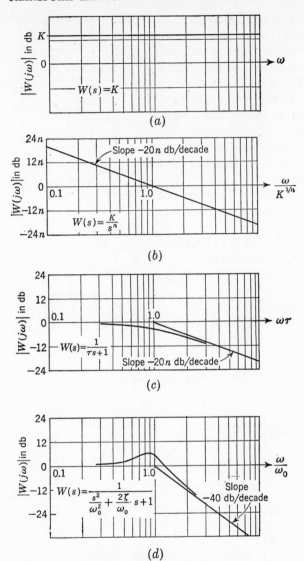

Fig. 9-22. Summary of amplitude-response curves.

Phase Response. The phase angle of the steady-state response of a system to a sinusoidal input is, from (9-54), the angle of $W(j\omega)$. Let us write $W(j\omega)$ for our general system function (9-56):

$$W(j\omega) = K \frac{(j\omega\tau_1 + 1)\left[2j\zeta_3 \dfrac{\omega}{\omega_3} + \left(1 - \dfrac{\omega^2}{\omega_3{}^2}\right) \right] \cdots}{(j\omega)^n (j\omega\tau_2 + 1)\left[2j\zeta_4 \dfrac{\omega}{\omega_4} + \left(1 - \dfrac{\omega^2}{\omega_4{}^2}\right) \right] \cdots} \qquad (9\text{-}67)$$

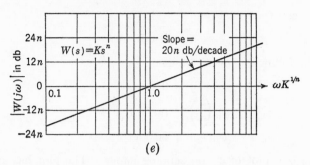

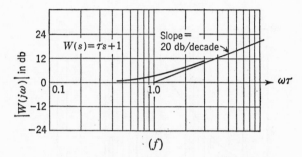

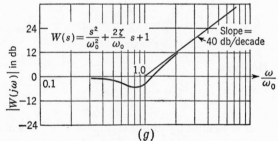

FIG. 9-22 (continued).

The angle ψ is

$$\psi = -n\frac{\pi}{2} + \tan^{-1}\omega\tau_1 + \tan^{-1}\frac{2\zeta_3\dfrac{\omega}{\omega_3}}{1 - \dfrac{\omega^2}{\omega_3{}^2}} + \cdots$$

$$- \tan^{-1}\omega\tau_2 - \tan^{-1}\frac{2\zeta_4\dfrac{\omega}{\omega_4}}{1 - \dfrac{\omega^2}{\omega_4{}^2}} - \cdots \quad (9\text{-}68)$$

The angle contributions of terms in the denominator are of the same form as those of the numerator, but with a sign change. We have two types of terms:

1. Let us call the first type

$$\psi_B(\omega) \triangleq \tan^{-1} \omega\tau$$

We note the following special values

$$\psi_B(0) = 0$$
$$\psi_B\left(\frac{1}{\tau}\right) = \tan^{-1} 1 = 45° \qquad (9\text{-}69)$$
$$\psi_B(\infty) = 90°$$

Figure 9-23 is a plot of ψ_B on semilog paper. The plot has symmetry about $\omega\tau = 1$.

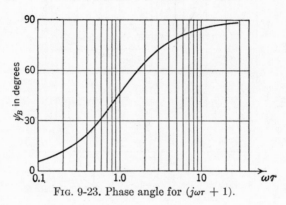

FIG. 9-23. Phase angle for $(j\omega\tau + 1)$.

2. Terms of the type

$$\psi_C(\omega) \triangleq \tan^{-1} \frac{2\zeta(\omega/\omega_0)}{1 - (\omega/\omega_0)^2} \qquad (9\text{-}70)$$

have the same kind of symmetry about $\omega/\omega_0 = 1$ as ψ_B has about $\omega\tau = 1$. Special values are

$$\psi_C(0) = 0$$
$$\psi_C(\omega_0) = 90° \qquad (9\text{-}71)$$
$$\psi_C(\infty) = 180°$$

A family of curves for various values of ζ is shown in Fig. 9-24.

Example. Let us plot amplitude and phase response for the transfer function

$$W(s) = \frac{10(s + 4)}{s(s^2 + s + 4)(s + 6)} \qquad (9\text{-}72)$$

We first put $W(s)$ in the form (9-56) for plotting:

$$W(s) = \frac{(s/4) + 1}{0.6s\left[\left(\dfrac{s}{2}\right)^2 + 2\left(\dfrac{0.25}{2}\right)s + 1\right]\left(\dfrac{s}{6} + 1\right)} \qquad (9\text{-}73)$$

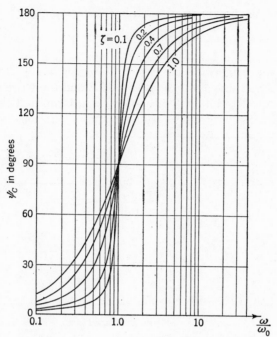

FIG. 9-24. Phase angle ψ_C for $\left[2j\zeta \dfrac{\omega}{\omega_0} + \left(1 - \dfrac{\omega^2}{\omega_0^2} \right) \right]$.

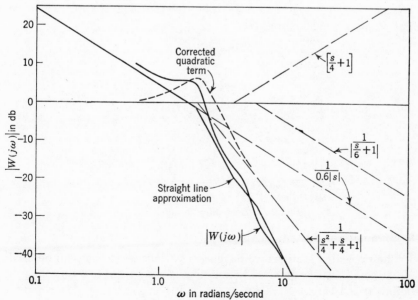

FIG. 9-25. Amplitude plot for (9-73).

The parameters of the quadratic term in the denominator are

$$\omega_0 = 2 \qquad (9\text{-}74)$$
$$\zeta = 0.25$$

Straight-line approximations are shown in Fig. 9-25. From Figs. 9-20 and 9-21 the peak value of 6.3 db at $\omega = 1.88$ and the value of 6.0 db at $\omega = 2$ have been determined for the quadratic term, and the corrected $|W(j\omega)|$ is compared with the straight-line approximation.

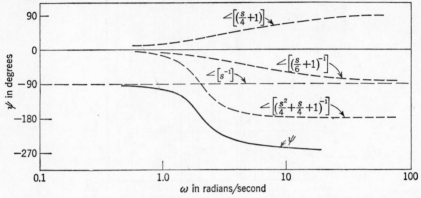

FIG. 9-26. Phase plot for (9-73).

In Fig. 9-26 the phase angle ψ is plotted from the angles of the factors making up $W(j\omega)$.

We shall use frequency-response plots in the next chapter when we discuss feedback-system analysis.

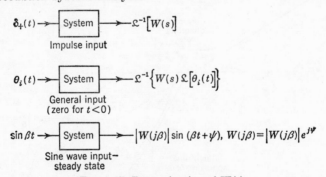

FIG. 9-27. Determination of $W(s)$.

9-6. Summary of System Descriptions

In the preceding sections we have seen various ways that can be used to describe a linear system. The relations among these descriptions will be summarized here.

If the system function $W(s)$ is known, we have a complete description of the linear system. We saw in Sec. 9-1 how $W(s)$ can be determined from the differential equations of the system. Figure 9-27 shows how it can be determined from experiments on the system.

The experiments in Fig. 9-27 are idealized in that we assume that the various inputs and outputs are known analytically. The problem of determining system function from experimental data may be

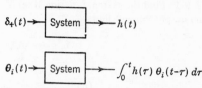

FIG. 9-28. Output determined from impulse response.

difficult. Various approximation techniques are available,[1] but we shall not go into this problem further here.

If the impulse response is measured, the output of the system in response to *any* input applied at $t = 0$ can be found without knowing the

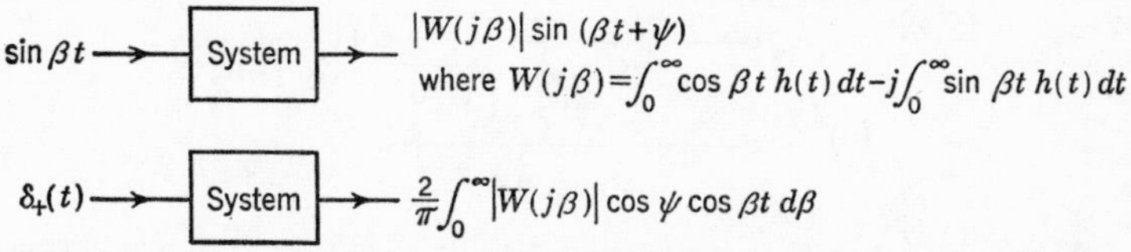

FIG. 9-29. Relations between impulse and frequency response.

analytic expressions for either the impulse response or the general input. The technique is summarized in Fig. 9-28. The integration can be carried out graphically.

Finally, the relations between impulse response and steady-state sine wave response are shown in Fig. 9-29. Table 9-1 summarizes the various relations among system descriptions.

TABLE 9-1. SUMMARY OF SYSTEM DESCRIPTIONS

	Assume These Are Known			
	System function $W(s)$	Impulse response $h(t)$	General output $\theta_0(t)$ (due to input $\theta_i(t)$, applied at $t = 0$)	Frequency response $W(j\beta)$
$W(s)$	—	$\mathcal{L}[h(t)]$	$\dfrac{\mathcal{L}[\theta_0(t)]}{\mathcal{L}[\theta_i(t)]}$	$[W(j\beta)]_{j\beta \to s}$
$h(t)$	$\mathcal{L}^{-1}[W(s)]$	—	From $W(s)$	$\dfrac{2}{\pi}\int_0^\infty |W(j\beta)|\cos\psi\cos\beta t\, d\beta$
$\theta_0(t)$	$\mathcal{L}^{-1}\{W(s)\mathcal{L}[\theta_i(t)]\}$	$\int_0^t h(\tau)\theta_i(t-\tau)\,d\tau$	—	From $W(s)$ or $h(t)$
$W(j\beta)$	$[W(s)]_{s\to j\beta}$	$\int_0^\infty \cos\beta t\, h(t)\,dt$ $-j\int_0^\infty \sin\beta t\, h(t)\,dt$	From $W(s)$	—

[1] See Truxal, *op. cit.*, chap. 6, for a detailed discussion of the determination of transfer functions from experimental data.

PROBLEMS

9-1. Find the system function $W(s) \triangleq \mathcal{L}[v_2]/\mathcal{L}[v_1]$ for the networks shown.   Assume no output current flows.

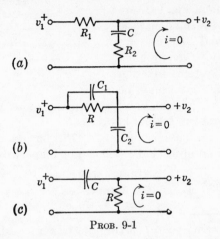

(a)

(b)

(c)

PROB. 9-1

9-2. Find the system function $\mathcal{L}(x_2)/\mathcal{L}(x_1)$ for the mechanical systems shown.

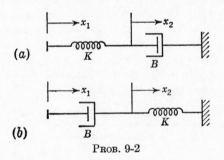

(a)

(b)

PROB. 9-2

9-3. Find $W(s)$ and impulse response $h(t)$.

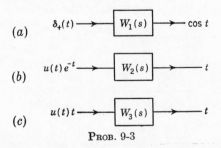

(a) $\delta_+(t) \longrightarrow \boxed{W_1(s)} \longrightarrow \cos t$

(b) $u(t)e^{-t} \longrightarrow \boxed{W_2(s)} \longrightarrow t$

(c) $u(t)t \longrightarrow \boxed{W_3(s)} \longrightarrow t$

PROB. 9-3

9-4. Show that, if initial conditions are zero, the unit step response of a system is the integral of the impulse response.

9-5. Make a sketch of frequency response for the systems in Prob. 9-3.

9-6. Given that $W(j\omega) = 1/(j\omega\tau + 1)$, use (9-53) to find $h(t)$.

9-7. Given that $h(t) = e^{-\alpha t}$, use (9-45) to find $W(j\beta)$.

9-8. Show that in a linear system with response $\theta_0(t)$ to $\theta_i(t)$ applied at $t = 0$, the response to $(d^n/dt^n)\theta_i(t)$ is $(d^n/dt^n)\theta_0(t)$.

9-9. Sketch block diagrams for the three types of gyros in Sec. 6-3.

9-10. Show that if $W(j\beta) = |W(j\beta)|e^{j\psi}$, ψ is an odd function of β and $\cos\psi$ is even, thus verifying (9-53).

CHAPTER 10

FEEDBACK-SYSTEM ANALYSIS

Linear feedback systems are widely used for a variety of engineering purposes. Automatic controls of aircraft, machines, and processes are examples of the applications of feedback devices. Our purpose here is to show how the system function and s-plane geometry can be used to analyze this kind of system.

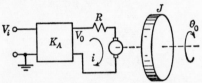

FIG. 10-1. Motor used to position load.

10-1. Linear Feedback Systems

Let us suppose that we wish to position a load of moment of inertia J with a d-c motor as shown in Fig. 10-1. We assume that the motor has constant field excitation and armature resistance R. An amplifier with voltage gain K_A drives the armature circuit. The equations describing the system are

$$
\begin{aligned}
V_0 &= K_A V_i \\
Ri + K_e \dot{\theta} &= V_0 \\
T &= K_T i \\
J \ddot{\theta} &= T
\end{aligned}
\tag{10-1}
$$

where K_e = back emf constant of motor

T = torque

J = moment of inertia

K_T = torque constant of motor

Combining Eqs. (10-1) and transforming, we have

$$
\begin{aligned}
\bar{V}_0(s) &= K_A \bar{V}_i(s) \\
\bar{\theta}_0(s) &= \frac{1}{K_e s (\tau s + 1)} \bar{V}_0(s) \qquad \tau \triangleq \frac{JR}{K_T K_e}
\end{aligned}
\tag{10-2}
$$

The system can be represented by a block diagram as shown in Fig. 10-2. This system would be a poor positioning device, since θ_0 will change linearly with time as long as any input is present because of the $1/s$ term in the motor/load transfer function. Skillful manipulation of V_i would be necessary to bring θ_0 to a desired position. Let us, therefore, change the system so that V_i is proportional to the error between desired and actual load positions. One such modification is shown in Fig. 10-3. We assume that the amplifier is capable of summing several inputs.

In the modified system a potentiometer has been added which provides K_1 volts/radian at the load. This voltage is subtracted from V_i,

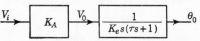

Fig. 10-2. Block diagram of amplifier-motor combination.

which now represents the desired load position and is derived from a potentiometer also. If the two voltages are equal, there is no error and the system remains at rest. If the actual and desired positions are not equal, the motor will move the load in a direction which will reduce the error. The block diagram of the new system is shown in Fig. 10-4. We have simplified the block diagram, using the techniques discussed in Sec. 9-2.

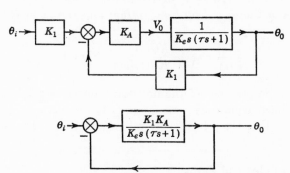

Fig. 10-3. Feedback added to Fig. 10-1.

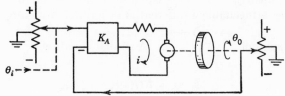

Fig. 10-4. Block diagram for Fig. 10-3.

We now ask if the system of Fig. 10-4 will do its positioning job satisfactorily. How, for example, will the load respond to a step function applied at θ_i? We shall return to this problem at the end of the chapter after we have developed some special techniques for the analysis of systems of this type.

Since the systems to be discussed in this chapter are linear, the methods discussed already are applicable to their analysis. However, a number of special techniques have been developed for feedback-system analysis, and it is the purpose of this chapter to describe these and the reasons for their use.

Figure 10-5 is a block diagram of a general feedback system. The arrows indicate the direction of signal flow and imply the presence of some sort of unilateral device in the loop.[1] The single direction of flow is characteristic of feedback devices and forms the only real difference between them and the other linear systems we have studied in previous chapters.

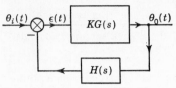

FIG. 10-5. General feedback system.

One equation for the system of Fig. 10-5 is

$$\bar{\epsilon}(s) = \bar{\theta}_i(s) - H(s)\bar{\theta}_0(s) \tag{10-3}$$

where $\bar{\theta}_i(s) = \mathcal{L}[\theta_i(t)]$, etc. This equation results from the subtracting device (such as a mechanical differential or a differential amplifier) indicated by the symbol ⊗ . A second equation is

$$\bar{\theta}_0(s) = KG(s)\bar{\epsilon}(s) \tag{10-4}$$

Combining (10-3) and (10-4) we derive an expression for the system function in terms of the *open-loop transfer functions* $G(s)$ and $H(s)$ and the *loop gain K:*

$$\frac{\bar{\theta}_0(s)}{\bar{\theta}_i(s)} = \frac{KG(s)}{1 + KG(s)H(s)} \triangleq Y(s) \tag{10-5}$$

We note that the presence of the unilateral device in the loop does not result in nonlinear equations. Equation (10-5) is the most general form of transfer function for a feedback system and forms the basis for our analysis.

10-2. System Poles and Zeros

Let us assume that the poles and zeros of $G(s)$ and $H(s)$ are known. We wish to know the poles and zeros of (10-5), since these will determine the behavior of the feedback system. The ways in which $Y(s)$ can have *zeros* are:

1. $Y(s)$ has a zero where $G(s)$ has a zero.

2. $Y(s)$ has a zero where $H(s)$ has a pole not canceled by a zero of $G(s)$.

The ways in which $Y(s)$ can have *poles* are:

---

[1] A vacuum-tube amplifier is a good approximation to a unilateral device, since signals applied at the output (plate) do not appear appreciably at the input (grid).

1. $Y(s)$ has a pole where $G(s)$ has a pole and $G(s)H(s)$ is finite or zero or has a pole of lower order.

2. $Y(s)$ has a pole where $1 + KG(s)H(s) = 0$.

We see that the zeros of the closed-loop transfer function can be found by inspection of $G(s)$ and $H(s)$. The poles in case 1 can be found this way also, but not so in case 2. The solution of the equation

$$1 + KG(s)H(s) = 0 \tag{10-6}$$

in general involves factoring a high-order polynomial, since $G(s)H(s)$ is a rational fraction. Although this factoring problem can be solved in a straightforward way by numerical analysis, the nature of the system often becomes immersed in the algebra. For this reason special techniques for finding the roots of (10-6) have been developed, and in the

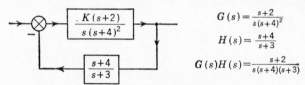

$$G(s) = \frac{s+2}{s(s+4)^2}$$

$$H(s) = \frac{s+4}{s+3}$$

$$G(s)H(s) = \frac{s+2}{s(s+4)(s+3)}$$

FIG. 10-6. Feedback system.

next section we shall discuss one of them—the *root-locus* method. All the special techniques for feedback-system analysis are based on methods for finding the roots of (10-6), preserving at the same time as much contact with the physics of the problem as possible.

Example. Before going on, we consider the system shown in Fig. 10-6 as an example. According to the discussion above, we can say immediately that

$Y(s)$ has zeros at $s = -2, -3$.
$Y(s)$ has pole at $s = -4$.

In addition, $Y(s)$ has poles at the roots of

$$1 + KG(s)H(s) = 1 + \frac{K(s+2)}{s(s+4)(s+3)} = 0 \tag{10-7}$$

or
$$s^3 + 7s^2 + (12 + K)s + 2K = 0 \tag{10-8}$$

Now from (10-5) the closed-loop transfer function $Y(s)$ is

$$
\begin{aligned}
Y(s) &= \frac{K(s+2)(s+3)}{(s+4)[s(s+4)(s+3) + K(s+2)]} \\
&= \frac{K(s+2)(s+3)}{(s+4)[s^3 + 7s^2 + (12 + K)s + 2K]}
\end{aligned}
\tag{10-9}
$$

which confirms our observations and makes apparent the problem of factoring—in this case, a cubic. We shall return to this problem after we have developed methods for solving it in the following sections.

10-3. The Root-locus Method[1]

Any complex value of s which satisfies Eq. (10-6) is the position in the s plane of a pole of the closed-loop transfer function. To find such values of s we write (10-6) as two equations:

$$|G(s)H(s)| = \frac{1}{K} \tag{10-10}$$

and[2]

$$\angle[G(s)H(s)] = n\pi \qquad n \text{ odd} \tag{10-11}$$

The analytical solution of (10-10) and (10-11) would be no simpler than that for (10-6). There is, however, a simple graphical solution based on the nature of $G(s)H(s)$ as a rational fraction and the geometry of the s-plane plot as discussed in Sec. 7-6. Let us write

$$G(s)H(s) = \frac{(s - s_1)(s - s_2) \cdots (s - s_Z)}{(s - \hat{s}_1)(s - \hat{s}_2) \cdots (s - \hat{s}_P)} \tag{10-12}$$

Now if we plot the poles and zeros of the typical $G(s)H(s)$ given in Eq. (10-12), the geometry will be as shown in Fig. 10-7. The directed line

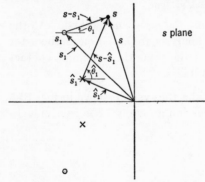

FIG. 10-7. s-plane geometry for the root-locus method.

segments $(s - s_1)$ and $(s - \hat{s}_1)$ have angles θ_1 and $\hat{\theta}_1$ and magnitudes given by the lengths of the lines. A trial point s can be tested as a solution of (10-6) by measuring the angles θ_1, $\hat{\theta}_1$, etc., and substituting their values in

$$\angle[G(s)H(s)] = (\theta_1 + \theta_2 + \cdots + \theta_Z) - (\hat{\theta}_1 + \hat{\theta}_2 + \cdots + \hat{\theta}_P)$$
$$= n\pi \qquad n \text{ odd} \tag{10-13}$$

where θ_1, θ_2, etc., are the angles of the lines drawn from the zeros of $G(s)H(s)$ to the trial point s and $\hat{\theta}_1$, $\hat{\theta}_2$, etc., are the angles of lines drawn from the poles.

If (10-13) is not satisfied—that is, if the measured angles do not sum to $n\pi$ with n odd—a new point s is tried.

If (10-13) is satisfied, the trial point is a root of (10-6) and the value of K is given by (10-10). We find the value of K graphically by writing

$$|G(s)H(s)| = \frac{l_1 \cdot l_2 \cdot \cdots \cdot l_Z}{\hat{l}_1 \cdot \hat{l}_2 \cdot \cdots \cdot \hat{l}_P} = \frac{1}{K} \tag{10-14}$$

---

[1] The root-locus method was developed by W. R. Evans. A detailed exposition is given by J. G. Truxal, "Automatic Feedback Control System Synthesis," chap. 4, McGraw-Hill Book Company, Inc., New York, 1955.

[2] $\angle[z]$ is the angle of the complex number z.

where l_1, l_2, etc., are the lengths of the lines drawn from the zeros of $G(s)H(s)$ to the point s, and l_1, l_2, etc., are the lengths of lines drawn from the poles.

Generally we find a locus of values of s satisfying the angle condition by (10-13) first. This is called the root locus. The value of K corresponding to any point on the locus can then be found from (10-14). A ruler and a protractor are the only tools necessary for the determination of the root locus.

10-4. Plotting the Locus

Before proceeding, it is well to have clearly in mind what the root locus is:

The root locus is the locus of roots of the equation $1 + KG(s)H(s) = 0$ as K varies from 0 to ∞.

These roots, for a given value of K, will be poles of the closed-loop transfer function $Y(s)$ of (10-5). *But* they are not the only poles. Others can be found directly from $G(s)$ and $H(s)$ as explained in Sec. 10-2. The closed-loop zeros, also found from $G(s)$ and $H(s)$, complete the pole-zero description of the closed-loop system. The root-locus method locates those poles of $Y(s)$ which cannot be found by inspection.

It is possible in most cases to sketch an approximate root-locus plot by following a set of simple rules. An accurate plot can always be made by measuring angles with a protractor according to (10-13).

In many problems we do not need the precise location of closed-loop poles and zeros. Often their approximate location is all that is required to make an estimate

Fig. 10-8. Real axis locus.

of system performance. If a more exact location is needed, the approximate plot is useful as a first step in its determination.

The rules for sketching the root-locus plot are given here. Proofs will be given in the next section.

1. *A locus always exists on the real axis when the sum of real poles and real zeros of $G(s)H(s)$ to the right is odd.*

Figure 10-8 illustrates the application of this rule for

$$G(s) = \frac{1}{s(s+1)(s+2)} \qquad H(s) = 1 \qquad (10\text{-}15)$$

2. *Far from the origin the loci are straight lines passing through the centroid of the poles and zeros of $G(s)H(s)$ with poles taken as positive masses and zeros as negative masses.   The centroid is on the real axis with distance from the origin given by*

$$\frac{\Sigma \ (real \ parts \ of \ poles) \ - \ \Sigma \ (real \ parts \ of \ zeros)}{P \ - \ Z} \tag{10-16}$$

The angles of the straight lines are given by

$$\phi = \frac{n\pi}{P - Z} \qquad n \ odd \tag{10-17}$$

P = number of finite poles of $G(s)H(s)$
Z = number of finite zeros of $G(s)H(s)$

In our example of Fig. 10-8 the centroid is at $(0 - 1 - 2)/3 = -1$. The angles are $\pi/3$, π, $5\pi/3$.   The next step in construction of the locus is shown in Fig. 10-9.

3. *The locus starts $(K = 0)$ on the poles of $G(s)H(s)$ and ends $(K = \infty)$ on the zeros of $G(s)H(s)$.*

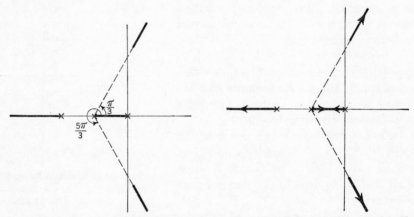

FIG. 10-9. Asymptotic loci.   FIG. 10-10. Motion of roots with K.

The motion of the roots with increasing K is shown by arrows.   In our example the zeros are at ∞, and the motion is shown in Fig. 10-10.

4. *The loci are continuous with K and have symmetry about the real axis.*

This rule enables us to finish the locus of Fig. 10-10.   Evidently the loci starting at $s = 0$ and $s = -1$ meet and leave the real axis to join the asymptotes.   The complete root-locus plot is shown in Fig. 10-11.

From the sketch of Fig. 10-11 a number of characteristics of the closed-loop system, which is shown in Fig. 10-12, can be inferred:

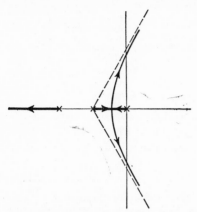

1. When the gain K is low, all the closed-loop poles are on the real axis and the impulse response is nonoscillatory.

2. As the gain is increased, a pair of poles moves off the real axis and the impulse response becomes oscillatory.

3. Upon further increase in gain the pair of poles moves into the right half plane and the impulse response

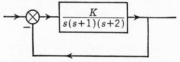

FIG. 10-11. Complete root-locus plot.   FIG. 10-12. Closed-loop system.

grows with time.   The system, then, becomes *unstable* if the gain is too high.

Other rules are available[1] giving, for example, the point of departure of the locus from the real axis, the angle of departure of a locus from a pole, etc.   However, the four rules given here will usually lead to a

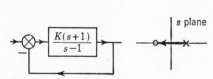

FIG. 10-13. Closed-loop system and root locus.

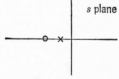

FIG. 10-14. Closed-loop pole-zero plot for Fig. 10-13.

sketch with sufficient detail to permit the kind of rough analysis of behavior given for the system in Fig. 10-12.

We shall illustrate the rules by the following examples.

Example 1.   Figure 10-13 shows a closed-loop system and its root-locus plot.   This system is unstable for low values of gain K.   If the gain is increased enough, the closed-loop pole moves into the left half plane and the closed-loop pole-zero plot might be as shown in Fig. 10-14.

Example 2.   In Fig. 10-15 a loop is closed around a system with a second-order pole at the origin.   The order of the pole is indicated by the number in parentheses and is treated like simple poles superimposed.

[1] Truxal, *loc. cit.*

In this case the asymptotes are at $\pm\pi/2$ and pass through the origin. The locus simply follows the asymptotes. As the gain is increased, the

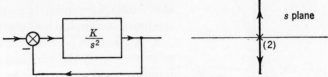

FIG. 10-15. Closed-loop system and root locus.

system impulse response is an undamped sinusoid of increasing frequency. A closed-loop pole-zero plot is shown in Fig. 10-16.

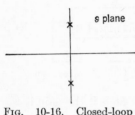

FIG. 10-16. Closed-loop pole-zero plot for Fig. 10-15.

Example 3. A closed-loop system and its root-locus plot are shown in Fig. 10-17.

This system becomes increasingly stable as K is increased, and for sufficiently large values of K the impulse response is nonoscillatory. An intermediate value of K would produce complex roots, and the closed-loop poles and zeros might be as shown in Fig. 10-18. We have included the $G(s)$ zero at $s = -3$ in accordance with the discussion in Sec. 10-2.

Example 4. Another closed-loop system and its root locus are shown in Fig. 10-19. In this case the asymptotes are at $\pm\pi/2$ and pass through

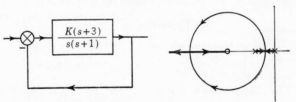

FIG. 10-17. Closed-loop system and root locus.

$s = [-2 - (-1)]/2 = -0.5$. The system is always oscillatory. As $K \to \infty$, the sinusoidal term will be multiplied by a damping term $\exp(-t/2)$, so that the system is always stable. Closed-loop poles and zeros for some intermediate value of K are shown in Fig. 10-20.

Example 5. The system of Fig. 10-6 is shown in Fig. 10-21 with its root-locus plot. Here the asymptotes are at $\pm\pi/2$ and pass through

$$s = (-4 - 3 + 2)/2 = -\tfrac{5}{2}.$$

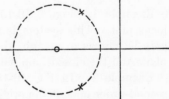

FIG. 10-18. Closed-loop pole-zero plot for Fig. 10-17.

Closed-loop poles and zeros for some K are shown in Fig. 10-22.

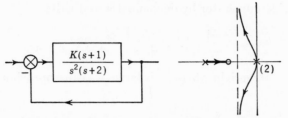

Fig. 10-19. Closed-loop system and root locus.

In this case the system is stable for all values of K. As K increases, the impulse response frequency of oscillation increases, the sinusoidal term being multiplied by exp $(-2.5t)$ for $K \to \infty$.

10-5. Proofs of the Root-locus Rules

The proofs of the rules stated in Sec. 10.4 will be given here.

Rule 1: *Real Axis Loci.* A typical open-loop set of poles and zeros is shown in Fig. 10-23. We see that the angle contributed at a point on the real axis by a pair of poles off the axis is always 2π, and that each pole and zero to the right of the point s contributes π. The ones to the left contribute zero. We conclude that if the number of real poles and zeros

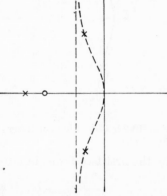

Fig. 10-20. Closed-loop poles and zeros of Fig. 10-19.

to the right of s is odd, then the angle of $G(s)H(s)$ is an odd multiple of π and the point in question is on a locus.

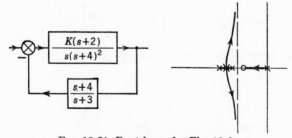

Fig. 10-21. Root locus for Fig. 10-6.

Rule 2: *Asymptotes.* We have written a general expression for $G(s)H(s)$ in (10-12). It is

$$G(s)H(s) = \frac{(s - s_1)(s - s_2) \cdots (s - s_Z)}{(s - \hat{s}_1)(s - \hat{s}_2) \cdots (s - \hat{s}_P)}$$

$$= \frac{s^Z - (s_1 + s_2 + \cdots + s_Z)s^{Z-1} + \cdots}{s^P - (\hat{s}_1 + \hat{s}_2 + \cdots + \hat{s}_P)s^{P-1} + \cdots} \quad (10\text{-}18)$$

We now divide numerator by denominator and write

$$G(s)H(s) = s^{Z-P} - [(s_1 + s_2 + \cdots + s_Z) \\ - (\hat{s}_1 + \hat{s}_2 + \cdots + \hat{s}_P)]s^{Z-P-1} + \cdots \quad (10\text{-}19)$$

Now far from the origin where $|s|$ is large, we can approximate $G(s)H(s)$

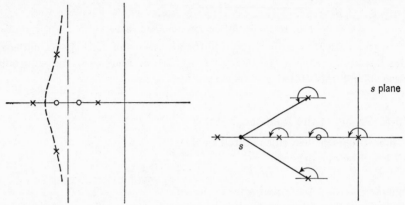

FIG. 10-22. Closed-loop pole-zero plot for   FIG. 10-23. Real axis locus angles.
Fig. 10-21.

by the first two terms in (10-19). The terms

$$(s_1 + s_2 + \cdots + s_Z) \triangleq z_r \qquad (10\text{-}20)$$
$$(\hat{s}_1 + \hat{s}_2 + \cdots + \hat{s}_P) \triangleq p_r$$

are the sums of the real parts[1] of the zeros and of the poles, respectively. Our problem is to solve (10-6) written for large $|s|$ in the form[2]

$$G(s)H(s) \cong s^{Z-P}\left[1 - \frac{z_r - p_r}{s}\right] \cong -\frac{1}{K} \qquad (10\text{-}21)$$

We now raise both sides to the $1/(Z-P)$ power:

$$s\left[1 - \frac{z_r - p_r}{s}\right]^{1/(Z-P)} = \left(-\frac{1}{K}\right)^{1/(Z-P)} \qquad (10\text{-}22)$$

and since $|s|$ is large, we can write (10-22)

$$s\left[1 - \frac{z_r - p_r}{(Z-P)s}\right] \cong \left(-\frac{1}{K}\right)^{1/(Z-P)} \qquad (10\text{-}23)$$

[1] If, for example, s_1 and s_2 are a conjugate pair $s_1 = \alpha + j\beta$, $s_2 = \alpha - j\beta$, then $s_1 + s_2 = 2\alpha$, the sum of real parts.

[2] This approach was suggested by H. Lass, A Note on the Root Locus Method, *Proc. IRE*, **44**:693 (May, 1956).

so that, finally, the solution is

$$s \cong \frac{p_r - z_r}{(P - Z)} + \left(-\frac{1}{K}\right)^{1/(Z-P)}$$

$$= \frac{p_r - z_r}{P - Z} + (K^{1/(P-Z)})(-1)^{1/(Z-P)} \tag{10-24}$$

Writing $-1 = \exp(-jn\pi)$, n odd, we have

$$s \cong \frac{p_r - z_r}{P - Z} + (K^{1/(P-Z)})e^{j[n\pi/(P-Z)]} \qquad n \text{ odd} \tag{10-25}$$

The solution s is the sum of two vectors shown in Fig. 10-24. For large values of $|s|$, then, the locus will vary with K along a straight line with the angle shown which passes through the real axis at the point given by Rule 2.

Rule 3: *Behavior with K.* The magnitude condition (10-10):

$$|G(s)H(s)| = \frac{1}{K} \tag{10-26}$$

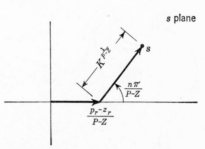

FIG. 10-24. Solution of $1 + KG(s)H(s) = 0$ for large $|s|$.

shows the behavior of the locus with K. When $K \to 0$, $|G(s)H(s)| \to \infty$, which can happen only at a pole of $G(s)H(s)$. When $K \to \infty$, $G(s)H(s) \to 0$, which happens only at the zeros of $G(s)H(s)$. Therefore the locus starts for $K = 0$ on the poles of $G(s)H(s)$ and ends for $K \to \infty$ on the zeros.

Rule 4: *Continuity and Symmetry.* The gain K can be written from (10-14) as

$$K = \frac{l_1 \cdot l_2 \cdots l_P}{l_1 \cdot l_2 \cdots l_Z} \tag{10-27}$$

where l_1 is the distance from a point s on the locus to a pole at $\hat{s}_1$ and l_1 is the distance to a zero at $s = s_1$. A small change in K can be accounted for by small changes in the lengths, corresponding to a continuous motion of s along the locus as K changes continuously.

The root locus represents possible closed-loop poles. Since the system function must be a ratio of polynomials in s with real coefficients,[1] the poles must occur as complex conjugate pairs. Since conjugate poles have symmetry about the real axis, the root locus must have the same symmetry.

[1] See Sec. 7-6.

10-6. The Nyquist Method

When the poles and zeros of $G(s)$ and $H(s)$ are known, the root-locus method can be used to find the closed-loop pole locations for a given gain K. Sometimes, however, $G(s)$ and $H(s)$ are not known but must be determined from experimental data. In this case the analysis problem includes the generally difficult job of finding analytic forms for $G(s)$ and $H(s)$ from measured frequency- or transient-response data.

A method is available which can be used to establish the degree of stability of a closed-loop system directly from frequency-response data— the Nyquist method. Although the Nyquist method does not locate closed-loop poles exactly—as does root locus—if often provides information about closed-loop-system behavior which could be obtained by root locus only after considerable labor.

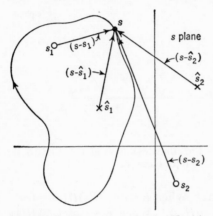

FIG. 10-25. s-plane geometry for Nyquist method.

The principle behind the Nyquist method is a geometric one. Figure 10-25 shows a pole-zero configuration like that we used in Fig. 10-7 to derive root locus. This time, however, we shall let s be a point on a closed path or contour in the s plane. A pole and a zero outside the closed path have been added. We now note that as the point s moves once around the path in the clockwise direction shown by the arrow, the following things happen:

1. The terms $(s - \hat{s}_1)$ and $(s - s_1)$, which correspond to poles and zeros *inside* the contour, change their angles clockwise by 2π each.

2. The terms $(s - \hat{s}_2)$ and $(s - s_2)$ belonging to poles and zeros *outside* the contour have each a net angle change of zero.

Now let us observe the effect this motion of s has on a rational fraction $F(s)$:

$$F(s) = \frac{(s - s_1)(s - s_2) \cdots}{(s - \hat{s}_1)(s - \hat{s}_2) \cdots} \tag{10-28}$$

When a term in the numerator changes by 2π, so does $F(s)$; when a term in the denominator changes by 2π, $F(s)$ changes by -2π. We can now state a general property:

A function $F(s)$ having P poles and Z zeros inside a closed contour will increase its angle clockwise by $2\pi(Z - P)$ as s moves once clockwise around the contour.

Evidently if we allow s to go clockwise around a path in the s plane, the difference between the number of zeros and poles of $F(s)$ inside the contour can be found by measuring the angular change in $F(s)$. In particular, if we let[1]

$$F(s) = \frac{1}{K} + G(s)H(s) \tag{10-29}$$

and choose a contour for s which encloses the right half plane, we can determine something about closed-loop-system stability.

Now the easiest way to count the number of times $F(s)$ changes by 2π is to plot $F(s)$ as a point in a complex plane and count the number of times this point encircles the origin. The $F(s)$ plane is shown in Fig. 10-26.

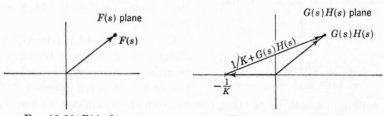

Fig. 10-26. $F(s)$ plane.   Fig. 10-27. $G(s)H(s)$ plane.

Since our interest is in $F(s)$ given by (10-29), we can make a plot in the $G(s)H(s)$ plane of Fig. 10-27 and count encirclements about the $-(1/K)$ point. We can now summarize the general Nyquist method.

1. An area of interest is chosen in the s plane.

2. Values of s along the path enclosing this area are substituted in $G(s)H(s)$.

3. The path traced by $G(s)H(s)$ is plotted in the $G(s)H(s)$ plane.

4. The number of clockwise encirclements of the $-(1/K)$ point of the $G(s)H(s)$ plane will be equal to $Z - P$ of $1/K + G(s)H(s)$ inside the s-plane contour.

5. The poles of $1/K + G(s)H(s)$ are the same as those of $G(s)H(s)$ which were assumed known in 2. Therefore the number P is known and

$$Z = P + [\text{No. of clockwise encirclements of } -(1/K)] \tag{10-30}$$

6. The number Z is the number of poles of the closed-loop-system function inside a given s-plane contour.

The method outlined above requires the expressions for $G(s)$ and $H(s)$ as does root locus, but provides less information about system behavior.

[1] The roots of $1 + KG(s)H(s) = 0$ are, of course, the same as those of

$$\frac{1}{K} + G(s)H(s) = 0$$

It is somewhat more convenient to use the latter form in this discussion.

We can, however, choose an s-plane path which will make possible the application of the method when only the measured frequency responses of the system components are known.

10-7. The Nyquist Diagram

Let us choose as the area of interest in the s plane the entire right half plane. Our contour is shown in Fig. 10-28. We have shown typical

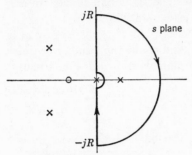

$G(s)H(s)$ poles and zeros in the figure. If a pole occurs on the imaginary axis, the contour is deformed into a small semicircle to exclude it from the interior of the contour. To enclose the whole right half plane we let $R \to \infty$.

If we now substitute the values of s along this contour in $G(s)H(s)$, the resulting plot in the $G(s)H(s)$ plane is known as a *Nyquist diagram*.

FIG. 10-28. s-plane contour.

A simplification in plotting results from the symmetry of the s-plane contour about the real axis. This means that for each value of s on the upper half of the contour there is also a complex conjugate value s^* on the lower half. Since $G(s)H(s)$ and $G(s^*)H(s^*)$ are complex conjugates, we need plot $G(s)H(s)$ for values of s in the upper half plane only. The remainder of the $G(s)H(s)$ plot, being the complex conjugate of the part already plotted, can be filled in so that the finished plot has the required real-axis symmetry.

Now let us examine the behavior of $G(s)H(s)$ as the large semicircle in Fig. 10-28 is traversed. The function $G(s)H(s)$ is assumed to be a rational fraction:

$$G(s)H(s) = \frac{(s - s_1)(s - s_2) \cdots (s - s_Z)}{(s - \hat{s}_1)(s - \hat{s}_2) \cdots (s - \hat{s}_P)}$$

$$= \frac{s^Z + \cdots}{s^P + \cdots} \tag{10-31}$$

For $s = R \exp (j\theta)$ on the large semicircle, $|s| = R$, a large number, so that

$$G(s)H(s) \cong \frac{R^Z e^{jZ\theta}}{R^P e^{jP\theta}} = R^{Z-P} e^{j(Z-P)\theta} \tag{10-32}$$

If, as is usually the case in practice, $P > Z$, we have, as $R \to \infty$,

$$G(s)H(s) \to 0 \tag{10-33}$$

and, when $P = Z$,

$$G(s)H(s) \to 1 \tag{10-34}$$

In neither case does the movement of s around the large semicircle cause any change in $G(s)H(s)$. This then means that the large semicircle in the s plane plots in the $G(s)H(s)$ plane as a single point, and we have the rule:

When the number of zeros of $G(s)H(s)$ does not exceed the number of poles, the large semicircle in the s plane need not be considered in plotting the Nyquist diagram.

We now have reduced the necessary part of the s-plane contour to the imaginary axis in the upper half plane. The value of s along this path is $j\omega$, so we need plot only $G(j\omega)H(j\omega)$—the frequency response.

One more thing needs attention—the small semicircle we made to avoid the pole at the origin in Fig. 10-28. Let us suppose that $G(s)H(s)$ has an nth-order pole at the origin. Then near the origin

$$G(s)H(s) \cong \frac{1}{s^n} \tag{10-35}$$

Now on the small semicircle of radius ϵ we can write

$$s = \epsilon e^{j\theta} \tag{10-36}$$

so that

$$G(s)H(s) \cong \frac{1}{\epsilon^n} e^{-jn\theta} \tag{10-37}$$

As $\epsilon \to 0$, the magnitude of $G(s)H(s)$ becomes large: as the point s changes its angle counterclockwise by π, the angle of $G(s)H(s)$ changes clockwise by $n\pi$. We say then:

The small semicircle in the s plane used to avoid an nth-order pole of $G(s)H(s)$ at the origin plots as a clockwise rotation of $n\pi$ at large magnitude in the $G(s)H(s)$ plane.

We shall illustrate the application of the Nyquist diagram with the same examples that were used for the root-locus method in Sec. 10-4.

In our examples the analytic form of $G(s)H(s)$ is known. We shall use the techniques of Sec. 9-5 to plot from this the magnitude and angle of $G(j\omega)H(j\omega)$. The polar plot obtained from the magnitude and angle plots is the Nyquist diagram. It should be kept in mind, however, that the polar plot could also be made from experimentally obtained magnitude and phase angle.

We have emphasized the usefulness of the Nyquist diagram in analyzing closed-loop systems from measured data. Although root locus provides more information about closed-loop behavior than does Nyquist when the poles and zeros of $G(s)$ and $H(s)$ are known, the Nyquist diagram is often used when root locus could be used also. There is a historical reason for this—Nyquist's method was published in 1932[1] and Evans's root-locus method in 1948.[2]   In addition, the Nyquist diagram may pro-

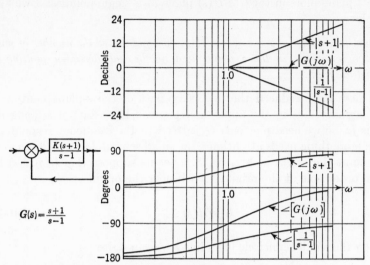

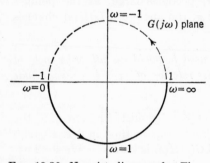

$$G(s) = \frac{s+1}{s-1}$$

FIG. 10-29. System and open-loop frequency response.

vide, for example, the gain at which roots cross into the right half plane more directly (see Example 1 below).   At any rate, the more methods of analysis we have for a problem, the better our insight into it.

Example 1. The closed-loop system and the open-loop frequency response with its components identified are shown in Fig. 10-29. The magnitude and phase of $G(j\omega)$ are used to make the polar plot in Fig. 10-30. The number of poles of the closed-loop system in the right half plane is equal to the number of zeros Z of $1/K + G(s)H(s)$ there.

FIG. 10-30. Nyquist diagram for Fig. 10-29.

The number Z is given by (10-30) to be

$$Z = P + \left[\text{No. of clockwise encirclements of } -\frac{1}{K} \right] \quad (10\text{-}38)$$

[1] H. Nyquist, Regeneration Theory, *Bell System Tech. J.*, January, 1932.

[2] W. R. Evans, Graphical Analysis of Control Systems, *Trans. AIEE*, **67**:547–551 (1948).

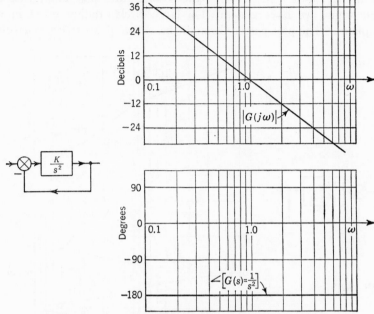

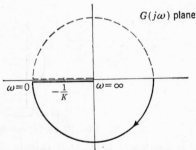

FIG. 10-31. System and open-loop frequency response.

From Fig. 10-29, P, the number of poles of $G(s)H(s)$ in the right half plane, is one. We now make the following observations from Fig. 10-30:

1. For $K < 1$ the number of encirclements is zero, and from (10-38)

$$Z = 1 \qquad (10\text{-}39)$$

That is, the closed-loop system has one pole in the right half plane.

2. For $K > 1$ there is one counterclockwise encirclement of the $-1/K$ point or -1 clockwise encirclements, and (10-38) gives

$$Z = 1 - 1 = 0 \qquad (10\text{-}40)$$

The closed-loop system has no poles in the right half plane.

3. For $K = 1$ the Nyquist plot passes through the $-1/K$ point. This corresponds to the boundary between right and left half planes, so that in this case the closed-loop system has a pole on the imaginary axis. Since there is only one pole, it must be at the origin.

FIG. 10-32. Nyquist plot for Fig. 10-31.

Example 2. Figure 10-31 shows the closed-loop system and its open-loop frequency response. The Nyquist diagram is plotted from Fig.

10-31 in Fig. 10-32. In avoiding the second-order pole of $G(s)$ at the s-plane origin, we have generated a large circle (radius $\rightarrow \infty$) in the $G(j\omega)$ plane. The plot passes through the $-1/K$ point twice, regardless

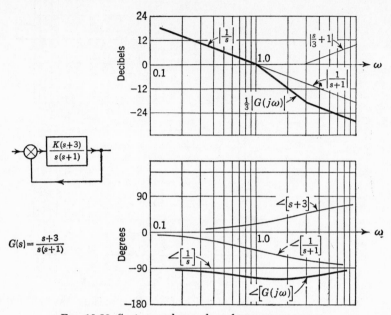

Fig. 10-33. System and open-loop frequency response.

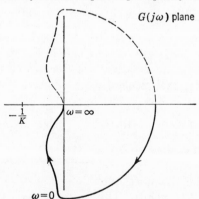

Fig. 10-34. Nyquist diagram for Fig. 10-33.

of the value of K, so that the closed-loop system always has two poles on the imaginary axis of the s plane.

Example 3. Figure 10-33 shows a closed-loop system and the open-loop frequency response. The Nyquist diagram is shown in Fig. 10-34.

In this case $P = 0$ and the number of encirclements is zero regardless of the value of K. The system is always stable.

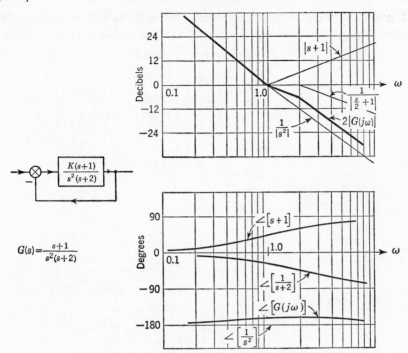

$$G(s) = \frac{s+1}{s^2(s+2)}$$

Fig. 10-35. System and open-loop frequency response.

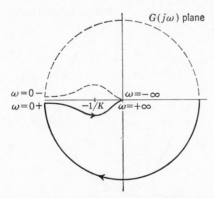

Fig. 10-36. Nyquist diagram for Fig. 10-35.

Example 4. Figure 10-35 shows another system and its open-loop frequency response. The Nyquist diagram is shown in Fig. 10-36.

As in Example 3, the number P is zero and so is the number of encirclements of $-1/K$. The system is always stable.

Example 5. The system of Fig. 10-6 and its open-loop frequency response are shown in Fig. 10-37. The Nyquist diagram is shown in Fig. 10-38.

Since the number of open-loop poles in the right half plane is zero, and since no encirclements of $-1/K$ are possible, the system is always stable.

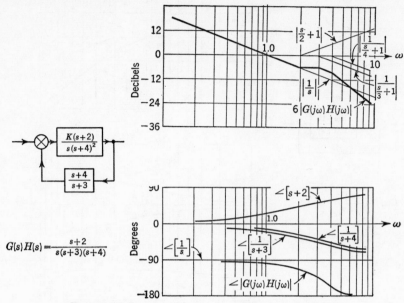

FIG. 10-37. System and open-loop frequency response.

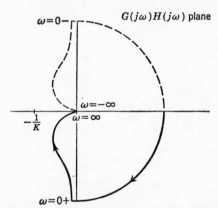

FIG. 10-38. Nyquist diagram for Fig. 10-37.

10-8. Improving System Performance

Let us now return to our simple position servomechanism of Figs. 10-3 and 10-4. The root-locus analysis is shown in Fig. 10-39. Evidently the system is always stable. With increasing loop gain, however, the system impulse response oscillates with increasing frequency and decreasing damping ratio. At first thought the solution would seem to be to keep the gain low. This cannot always be done, since it has an adverse effect on steady-state performance. Let us use the final-value theorem to examine the error under step and ramp inputs.

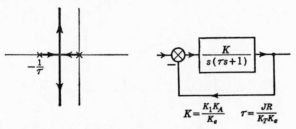

Fig. 10-39. Root locus for system of Fig. 10-4.

The error is defined here to be the difference between desired and actual output:

$$\epsilon(t) = \theta_i(t) - \theta_0(t) \tag{10-41}$$

Transforming, we have

$$\begin{aligned}\bar{\epsilon}(s) &= \bar{\theta}_i(s) - \bar{\theta}_0(s) \\ &= \bar{\theta}_i(s) - Y(s)\bar{\theta}_i(s)\end{aligned} \tag{10-42}$$

In the present case

$$Y(s) = \frac{KG(s)}{1 + KG(s)} \qquad G(s) = \frac{1}{s(\tau s + 1)} \tag{10-43}$$

so that we have

$$\begin{aligned}\bar{\epsilon}(s) &= \bar{\theta}_i(s)\left[1 - \frac{KG(s)}{1 + KG(s)}\right] \\ &= \frac{\bar{\theta}_i(s)}{1 + KG(s)} = \frac{s(\tau s + 1)}{s(\tau s + 1) + K}\bar{\theta}_i(s)\end{aligned} \tag{10-44}$$

Now, using the final-value theorem, we have for $\theta_i(t) = u(t)$, the unit step function,

$$\epsilon_1(\infty) = \left[\frac{s(\tau s + 1)}{s(\tau s + 1) + K}\right]_{s=0} = 0 \tag{10-45}$$

For $\theta_i(t) = t$

$$\epsilon_2(\infty) = \left[\frac{\tau s + 1}{s(\tau s + 1) + K}\right]_{s=0} = \frac{1}{K} \tag{10-46}$$

The error in the steady state with a step input is always zero. The error with a ramp input[1] varies as $1/K$ and will be reduced by using a large loop gain. Evidently, as the system stands, the requirements of good damping and low error with ramp input are inconsistent.

[1] Since the input was a ramp with unit slope, ϵ_2 is error per unit slope.

Suppose now that we could add a zero to the root-locus plot of Fig. 10-39. The new plot would look like Fig. 10-40. We can now increase the gain without decreasing damping. The addition of the zero to our system can be accomplished with a tachometer as shown in Fig. 10-41.

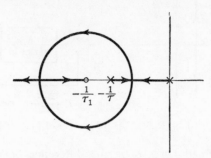

FIG. 10-40. Effect of adding a zero to plot of Fig. 10-39.

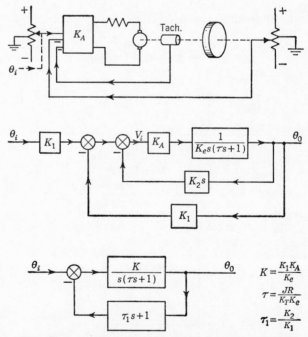

FIG. 10-41. System with tachometer feedback and block diagrams.

The system block diagram has been simplified as before. The tachometer provides K_2 volts/radian per sec. We have now established that a zero will help the impulse response damping and that a tachometer will provide the zero. Let us finally check the steady-state error. In this case

$$\bar{\epsilon}(s) = \bar{\theta}_i(s)\left[1 - \frac{KG(s)}{1 + KG(s)H(s)}\right] \qquad (10\text{-}47)$$

where
$$G(s) = \frac{1}{s(\tau s + 1)} \qquad (10\text{-}48)$$

$$H(s) = \tau_1 s + 1$$

so
$$\bar{\epsilon}(s) = \bar{\theta}_i(s)\left[\frac{1 + \dfrac{K(\tau_1 s + 1)}{s(\tau s + 1)} - \dfrac{K}{s(\tau s + 1)}}{1 + \dfrac{K(\tau_1 s + 1)}{s(\tau s + 1)}}\right]$$

$$= \bar{\theta}_i(s)\left[\frac{s(\tau s + 1 + K\tau_1)}{s(\tau s + 1) + K(\tau_1 s + 1)}\right] \qquad (10\text{-}49)$$

The steady-state error for a unit step input is

$$\epsilon_1(\infty) = \left[\frac{s(\tau s + 1 + K\tau_1)}{s(\tau s + 1) + K(\tau_1 s + 1)}\right]_{s=0} = 0 \qquad (10\text{-}50)$$

and for a unit ramp

$$\epsilon_2(\infty) = \left[\frac{\tau s + 1 + K\tau_1}{s(\tau s + 1) + K(\tau_1 s + 1)}\right]_{s=0} = \frac{1 + K\tau_1}{K} \qquad (10\text{-}51)$$

To study the effect of the parameter τ_1 (that is, the zero location) let us compare steady-state ramp input error and damping ratio. From Sec. 9-5 we compare the general second-order term

$$s^2 + 2\zeta\omega_0 s + \omega_0{}^2$$

with the denominator of (10-49) divided by τ:

$$s^2 + \frac{1 + K\tau_1}{\tau}s + \frac{K}{\tau}$$

and write

$$\zeta = \frac{1 + K\tau_1}{2\sqrt{K\tau}} = \frac{1 + (K\tau)(\tau_1/\tau)}{2\sqrt{K\tau}} \qquad (10\text{-}52)$$

in terms of the *dimensionless parameters* $K\tau$ and τ_1/τ. Also, from (10-51) we have

$$\epsilon_2(\infty) = \frac{1 + K\tau_1}{K} \qquad (10\text{-}53)$$

or
$$\frac{\epsilon_2}{\tau} = \frac{1}{K\tau} + \frac{\tau_1}{\tau} \qquad (10\text{-}54)$$

Combining (10-54) and (10-52) we have also

$$\zeta = \frac{1}{2}\frac{\epsilon_2}{\tau}\sqrt{K\tau} \qquad (10\text{-}55)$$

A plot of constant values of error and time constant ratio is shown in Fig. 10-42.

Suppose, for example, that in the original system ($\tau_1/\tau = 0$) of Fig. 10-39 we require ϵ_2/τ of 0.5. The corresponding damping ratio is 0.35. We see that the addition of the zero in Fig. 10-40 makes possible the same error with an improved ζ of 0.5 if $\tau_1/\tau = 0.25$ and the gain K is doubled.

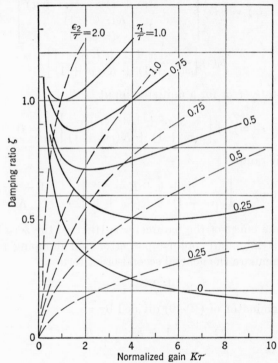

FIG. 10-42. Design chart for system of Fig. 10-41.

This example indicates one approach which can be taken in designing feedback systems. Our emphasis has been on the special techniques of analysis, using the ideas of system functions, block diagrams, and s-plane geometry from the preceding chapters. The material in this chapter forms the basis for feedback-system design. The reader is referred elsewhere for a more extensive discussion.[1]

PROBLEMS

10-1. Verify the block-diagram simplification in Figs. 10-4 and 10-41.

10-2. Derive a rule giving the angle of departure of the root locus from a pole.

[1] See, for example, Truxal, *op. cit.*

10-3. Sketch root loci for

a. $G(s) = \dfrac{1}{s^3}$, $H(s) = 1$

b. $G(s) = \dfrac{1}{s^3}$, $H(s) = (s + 1)$

c. $G(s) = \dfrac{1}{s(s^2 + 4)}$, $H(s) = 1$

d. $G(s) = \dfrac{1}{s(s^2 + 4)}$, $H(s) = (s + 1)$

e. $G(s) = \dfrac{1}{s^4}$, $H(s) = 1$

f. $G(s) = \dfrac{1}{s(s + 1)(s + 2)(s + 3)}$, $H(s) = 1$

g. $G(s) = \dfrac{1}{s[(s + 1)^2 + 1](s + 2)}$, $H(s) = 1$ (Be careful of this one!)

h. $G(s) = \dfrac{s + 1}{(s - 1)(s^2 + 4)}$, $H(s) = 1$

i. $G(s) = \dfrac{1}{s^2(s + 3)}$, $H(s) = \dfrac{s}{s + 2}$

j. $G(s) = 1$, $H(s) = \dfrac{s + 1}{s + 2}$

k. $G(s) = 1$, $H(s) = s$

l. $G(s) = s^3$, $H(s) = 1$

m. $G(s) = s^2$, $H(s) = \dfrac{1}{(s - 1)^2}$

10-4. Plot closed-loop poles and zeros of Prob. 10-3 for some value of K.

10-5. Sketch Nyquist diagrams for Prob. 10-3, and discuss stability.

10-6. Discuss the effect on the Nyquist method of avoiding a pole of $G(s)H(s)$ at the origin by detouring to the *left* (see Fig. 10-28).

10-7. Determine which of the systems in Prob. 10-3 may become unstable for large K. For these investigate the effects of adding poles and zeros to the loop. Find at least one method of stabilization for each.

10-8. Investigate steady-state error [defined as (*input*) − (*output*)] for step and ramp inputs of systems in Prob. 10-3. Which systems are unsatisfactory as positioning devices? Can you find other uses?

$\mathcal{L}$-TRANSFORM SOLUTION
OF PARTIAL DIFFERENTIAL EQUATIONS

We have already seen in Sec. 8-7 how the $\mathcal{L}$ transform can be used to solve partial differential equations. In this chapter we shall go into more detail, deriving the equations for problems in vibration, heat, and transmission lines and solving by the $\mathcal{L}$ method. In Chaps. 12 and 17 we shall show how other kinds of transforms can be applied to partial differential equations.

The subject of partial differential equations is a large one and an important one in the physical sciences. Our intent here is to show only how the $\mathcal{L}$ method can be applied to it.[1]

11-1. General Remarks

A partial differential equation always has two or more independent variables. We saw in Sec. 8-7 in connection with the equation

$$\frac{\partial^2 y}{\partial x^2} - \frac{1}{v^2} \frac{\partial^2 y}{\partial t^2} = 0 \qquad (11-1)$$

how the $\mathcal{L}$ transform replaced the space variable x by the transform variable s. The result was an ordinary differential equation in t, with s treated as a parameter.

In transforming (11-1) we made use of the fact that

$$\mathcal{L}_t \left[\frac{\partial^2 y(x,t)}{\partial x^2} \right] = \frac{\partial^2 Y_1(x,s)}{\partial x^2} = \frac{d^2 Y_1(x,s)}{dx^2} \qquad (11-2)$$

where $Y_1(x,s) \triangleq \mathcal{L}_t[y(x,t)]$. The ordinary derivative is written because there is only one variable x in $Y_1(x,s)$. The other term in (11-1) transformed into

$$\mathcal{L}_t \left[\frac{\partial^2 y(x,t)}{\partial t^2} \right] = s^2 Y_1(x,s) - sy(x,0) - y_t(x,0) \qquad (11-3)$$

[1] For additional material see, for example, I. N. Sneddon, "Fourier Transforms," McGraw-Hill Book Company, Inc., New York, 1951; R. V. Churchill, "Modern Operational Mathematics in Engineering," McGraw-Hill Book Company, Inc., New York, 1944.

The $\mathcal{L}$ transform of (11-1) is an ordinary differential equation:

$$\frac{d^2 Y_1(x,s)}{dx^2} - \frac{s^2}{v^2} Y_1(x,s) = \frac{1}{v^2} [sy(x,0) + y_t(x,0)] \tag{11-4}$$

The transform of a partial differential equation has one less independent variable than the original. Equations with more than two independent variables (e.g., time and three space coordinates) can be reduced to ordinary differential equations by repeated transforming. The resulting equation [like (11-4)] can itself be transformed in the usual way. It is important when successive transformations are made to choose a new transform variable each time.

Choice of the order in which independent variables are transformed generally depends on the nature of the problem. For example, we could have transformed (11-1) first with respect to x, and obtained

$$\frac{d^2 Y_2(\gamma,t)}{dt^2} - \gamma^2 v^2 Y_2(\gamma,t) = v^2[\gamma y(0,t) + y_x(0,t)] \tag{11-5}$$

where $\mathcal{L}_x[y(x,t)] \triangleq Y_2(\gamma,t)$. This form would be somewhat more convenient if, for example,

$$y(0,t) = y_x(0,t) = 0 \tag{11-6}$$

The transforms of the solutions of partial differential equations are more complicated than those for ordinary differential equations. As we shall see in the following sections, transcendental functions of s occur. A short table of such transforms will be found at the end of this chapter.

11-2. Vibrating String

Suppose we have a string of density ρ (mass/unit length) and under tension T. We assume that the deflection is small, so that the sines of angles between any segment of the string and the horizontal can be taken as equal to the angles.

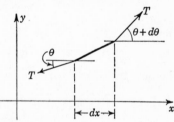

FIG. 11-1. Segment of a vibrating string.

Figure 11-1 shows a segment of such a string of length dx. The acceleration times the mass of this segment is equal to the sum of applied forces:

$$(\rho \, dx) \frac{\partial^2 y}{\partial t^2} = T(\theta + d\theta) - T\theta \tag{11-7}$$

The angle θ is the slope of the segment,

$$\theta = \frac{\partial y}{\partial x} \tag{11-8}$$

Then (11-7) becomes

$$\rho \, dx \, \frac{\partial^2 y}{\partial t^2} = T \, d\theta = T \, \frac{\partial \theta}{\partial x} \, dx = T \, \frac{\partial^2 y}{\partial x^2} \, dx \tag{11-9}$$

so that, finally,

$$\frac{\partial^2 y}{\partial x^2} - \frac{\rho}{T} \, \frac{\partial^2 y}{\partial t^2} = 0 \tag{11-10}$$

If we let

$$\frac{T}{\rho} \triangleq v^2 \tag{11-11}$$

we have in (11-10) an equation of the form of (11-1).

Example. Suppose the string is semi-infinite. We shall move one end with a prescribed motion $f(t)$ which begins at $t = 0$. Before that, the string is motionless. These conditions, together with the fact that the motion will always be zero at the far end of the string (at ∞), are called *boundary conditions*. There are as many conditions as there are derivatives in the equation—in this case, four:

$$y(0,t) = f(t) \tag{11-12}$$
$$y(\infty,t) = 0 \tag{11-13}$$
$$y(x,0) = 0 \tag{11-14}$$
$$y_t(x,0) = 0 \tag{11-15}$$

Conditions (11-14) and (11-15), being zero, suggest transforming first with respect to t, so that the transformed equation, from (11-4), is

$$\frac{d^2 Y(x,s)}{dx^2} - \frac{s^2}{v^2} \, Y(x,s) = 0 \tag{11-16}$$

where $\mathcal{L}_t[y(x,t)] \triangleq Y(x,s)$. We now transform again, this time letting

$$\mathcal{L}_x[Y(x,s)] \triangleq \hat{Y}(\gamma,s) \tag{11-17}$$

The new transform variable is γ. Equation (11-16) now becomes

$$\gamma^2 \hat{Y}(\gamma,s) - \frac{s^2}{v^2} \, \hat{Y}(\gamma,s) = \gamma Y(0,s) + Y_x(0,s) \tag{11-18}$$

Now from (11-12)

$$Y(0,s) = \mathcal{L}_t[y(0,t)] = \mathcal{L}_t[f(t)] \triangleq F(s) \tag{11-19}$$

The remaining condition (11-13), being a condition at ∞, is not immediately useful in evaluating $Y_x(0,s)$. We write its transform

$$\mathcal{L}_t[y(\infty,t)] = Y(\infty,s) = 0 \tag{11-20}$$

Equation (11-18) can be solved for $\hat{Y}(\gamma,s)$

$$\begin{aligned}
\hat{Y}(\gamma,s) &= \frac{\gamma F(s) + Y_x(0,s)}{\gamma^2 - (s^2/v^2)} \\
&= \frac{\frac{1}{2}[F(s) + (v/s)Y_x(0,s)]}{\gamma - (s/v)} + \frac{\frac{1}{2}[F(s) - (v/s)Y_x(0,s)]}{\gamma + (s/v)}
\end{aligned} \tag{11-21}$$

We now invert with respect to γ:

$$Y(x,s) = \mathcal{L}_\gamma^{-1}[\hat{Y}(\gamma,s)] = \frac{1}{2}\left[F(s) + \frac{v}{s}Y_x(0,s)\right]e^{(s/v)x}$$
$$+ \frac{1}{2}\left[F(s) - \frac{v}{s}Y_x(0,s)\right]e^{-(s/v)x} \quad (11\text{-}22)$$

But from (11-20), $Y(\infty,s)$ must be zero, so the coefficient of the first term on the right is zero:

$$F(s) + \frac{v}{s}Y_x(0,s) = 0$$
$$Y_x(0,s) = -\frac{s}{v}F(s) \quad (11\text{-}23)$$

and we have

$$Y(x,s) = F(s)e^{-(s/v)x} \quad (11\text{-}24)$$

We now invert (11-24), using (8-15):

$$y(x,t) = f\left(t - \frac{x}{v}\right)u\left(t - \frac{x}{v}\right) \quad (11\text{-}25)$$

We see that at a given time the disturbance represented by $f(t)$ has reached the point $x = vt$, so that v is properly interpreted as the velocity. A plot of deflection vs. time and distance is shown in Fig. 11-2 for

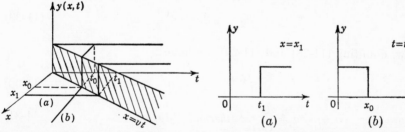

FIG. 11-2. Solution of vibrating string problem.

FIG. 11-3. Deflection of string as a function of t and x.

$f(t) = u(t)$. Slices have been cut across the solid of Fig. 11-2 to give curves a and b of Fig. 11-3. Curve a is a plot of deflection at point $x = x_1$ against time. We see that the disturbance does not reach x_1 until time t_1. Curve b is a "photograph" taken of the string at time t_0. It might be mentioned that the phenomenon can be studied experimentally with a garden hose.

11-3. Heat Conduction

Let us consider a slab of unit area and thickness dx as shown in Fig. 11-4. We adopt the following symbols:

$\theta(x,t)$ = temperature difference between a point in a body x and a reference point $x = 0$ at time t

q = quantity of heat per unit area per unit time

k = thermal conductivity

c = specific heat

It is an empirical fact that the amount of heat flowing at a point is proportional to the temperature gradient there. Then heat flow per unit time can be written

$$q = -k\frac{\partial \theta}{\partial x} \tag{11-26}$$

and

$$q + dq = -k\frac{\partial}{\partial x}(\theta + d\theta) = -k\frac{\partial}{\partial x}\left(\theta + \frac{\partial \theta}{\partial x}dx\right) \tag{11-27}$$

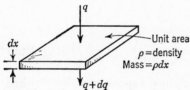

so that the net heat change per unit time in the slab is

$$dq = -k\frac{\partial^2 \theta}{\partial x^2}dx \tag{11-28}$$

FIG. 11-4. Heat flow in a slab.

Now the heat change per unit time is also given by the product of mass, specific heat, and rate of change of temperature:

$$-dq = \rho\,dx\cdot c\cdot\frac{\partial \theta}{\partial t} \tag{11-29}$$

Now, equating (11-28) and (11-29), we have

$$a\frac{\partial^2 \theta}{\partial x^2} - \frac{\partial \theta}{\partial t} = 0 \tag{11-30}$$

where

$$a \triangleq \frac{k}{c\rho} \qquad \textit{(thermal diffusivity)} \tag{11-31}$$

Example. Let us now solve a problem of heat conduction in a semi-infinite solid. Equation (11-30) requires three boundary conditions. These can be set as follows:

1. We shall assume that the solid has initially a uniform temperature: $\theta(x,0) = 0$.

2. At $t = 0$ the temperature of the face at $x = 0$ is raised to θ_0: $\theta(0,t) = \theta_0 u(t)$.

3. For very large values of x the temperature remains unchanged: $\theta(\infty,t) = 0$.

We now let

$$\mathcal{L}_t[\theta(x,t)] \triangleq \bar{\theta}(x,s) \tag{11-32}$$

The transform of (11-30) is

$$a \frac{d^2\bar{\theta}(x,s)}{dx^2} - s\bar{\theta}(x,s) + \theta(x,0) = 0 \qquad (11\text{-}33)$$

which becomes, because of condition (1),

$$a \frac{d^2\bar{\theta}(x,s)}{dx^2} - s\bar{\theta}(x,s) = 0 \qquad (11\text{-}34)$$

We now transform again, letting

$$\mathfrak{L}_x[\bar{\theta}(x,s)] \triangleq \hat{\theta}(\gamma,s) \qquad (11\text{-}35)$$

so that (11-34) becomes

$$a\gamma^2\hat{\theta}(\gamma,s) - a\gamma\bar{\theta}(0,s) - a\bar{\theta}_x(0,s) - s\hat{\theta}(\gamma,s) = 0 \qquad (11\text{-}36)$$

The transform of θ is obtained from (11-36)

$$\hat{\theta}(\gamma,s) = \frac{a\gamma\bar{\theta}(0,s) + a\bar{\theta}_x(0,s)}{a\gamma^2 - s}$$

From condition 2 we have

$$\bar{\theta}(0,s) = \mathfrak{L}_t[\theta(0,t)] = \frac{\theta_0}{s} \qquad (11\text{-}37)$$

As in the previous example, the evaluation of the remaining condition comes from the value of $\theta(\infty,t)$ in condition 3.   We now write

$$\hat{\theta}(\gamma,s) = \frac{(\gamma/s)\theta_0 + \bar{\theta}_x(0,s)}{\gamma^2 - (s/a)} \qquad (11\text{-}38)$$

$$= \frac{\frac{1}{2}\left[\frac{1}{s}\theta_0 + \sqrt{\frac{a}{s}}\,\bar{\theta}_x(0,s)\right]}{\gamma - \sqrt{s/a}} + \frac{\frac{1}{2}\left[\frac{1}{s}\theta_0 - \sqrt{\frac{a}{s}}\,\bar{\theta}_x(0,s)\right]}{\gamma + \sqrt{s/a}} \qquad (11\text{-}39)$$

Inverting, we have

$$\bar{\theta}(x,s) = \mathfrak{L}_\gamma^{-1}[\hat{\theta}(\gamma,s)] = \frac{1}{2}\left[\frac{1}{s}\theta_0 + \sqrt{\frac{a}{s}}\,\bar{\theta}_x(0,s)\right] e^{\sqrt{(s/a)}x}$$

$$+ \frac{1}{2}\left[\frac{1}{s}\theta_0 - \sqrt{\frac{a}{s}}\,\bar{\theta}_x(0,s)\right] e^{-\sqrt{(s/a)}x} \qquad (11\text{-}40)$$

Condition 3 makes the coefficient of the first term on the right zero:

$$\frac{1}{s}\theta_0 + \sqrt{\frac{a}{s}}\,\bar{\theta}_x(0,s) = 0$$

$$\bar{\theta}_x(0,s) = -\frac{1}{\sqrt{as}}\theta_0 \qquad (11\text{-}41)$$

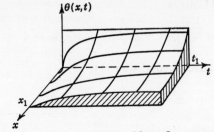

FIG. 11-5. Plot of solution of heat-flow example.

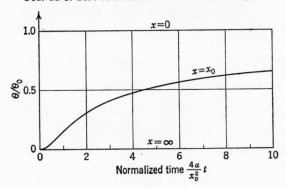

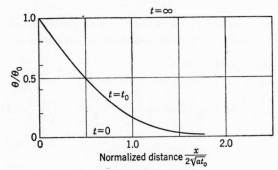

FIG. 11-6. Plots of $\theta(x,t)/\theta_0$.

So that we have from (11-40)

$$\bar{\theta}(x,s) = \theta_0 \frac{1}{s} e^{-\sqrt{(s/a)}x} \tag{11-42}$$

Referring to pair 7, Table 11-1, we have finally

$$\theta(x,t) = \theta_0 \operatorname{erfc}\left[\frac{x}{2\sqrt{at}}\right] \tag{11-43}$$

where erfc is the complementary error function discussed in Sec. 2-4. The solution can be plotted as a surface as in Fig. 11-5. The solution as a function of x and t separately is shown in Fig. 11-6.

11-4. Transmission Line

Unlike the networks studied in Chap. 4, the electrical parameters of a transmission line are distributed along the length of the line. Instead of having, say, a capacitor of so many farads, we have capacity per length of line. We shall consider here a line with no dissipation (no resistance). A section of length dx is shown in Fig. 11-7. The loop and node equations relating the quantities shown are

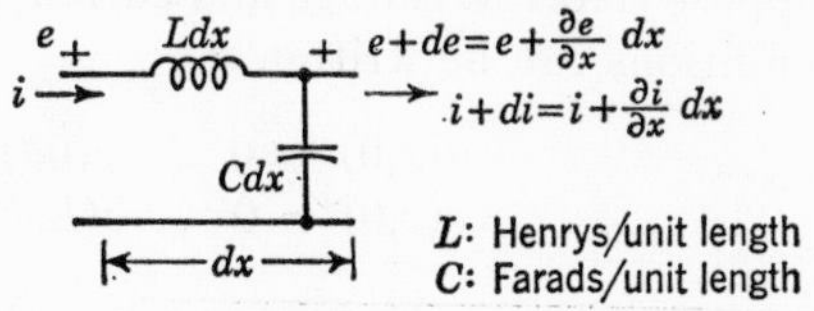

FIG. 11-7. Section of a transmission line.

$$e - L\,dx\,\frac{\partial i}{\partial t} = e + \frac{\partial e}{\partial x}\,dx$$
$$i = C\,dx\,\frac{\partial e}{\partial t} + i + \frac{\partial i}{\partial x}\,dx \tag{11-44}$$

which reduce to

$$L\frac{\partial i}{\partial t} + \frac{\partial e}{\partial x} = 0$$
$$C\frac{\partial e}{\partial t} + \frac{\partial i}{\partial x} = 0 \tag{11-45}$$

If we now differentiate the first of (11-45) $\partial/\partial t$ and the second $\partial/\partial x$ and combine, there results

$$LC\frac{\partial^2 i}{\partial t^2} - \frac{\partial^2 i}{\partial x^2} = 0 \tag{11-46}$$

Similarly, we can get, by differentiating first $\partial/\partial x$,

$$LC\frac{\partial^2 e}{\partial t^2} - \frac{\partial^2 e}{\partial x^2} = 0 \tag{11-47}$$

Now we could transform (11-46) or (11-47) and proceed as with the vibrating string, since the equations have the same form as (11-1). Let us instead work with (11-45). If we let

$$\mathcal{L}_t[i(x,t)] \triangleq I(x,s)$$
$$\mathcal{L}_t[e(x,t)] \triangleq E(x,s) \tag{11-48}$$

then Eqs. (11-45) become after transforming

$$LsI + \frac{dE}{dx} = Li(x,0)$$
$$CsE + \frac{dI}{dx} = Ce(x,0) \tag{11-49}$$

Example. Let us now fix boundary conditions. Suppose that we have a line of length L shorted at the far end as shown in Fig. 11-8. We further suppose that the voltage and current at any point are zero initially. The conditions can be written

$$i(x,0) = 0 \qquad e(0,t) = \epsilon(t)$$
$$e(x,0) = 0 \qquad e(L,t) = 0 \text{ (short circuit)} \tag{11-50}$$

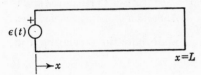

FIG. 11-8. Shorted transmission line.

Now we can write (11-49)

$$LsI + \frac{dE}{dx} = 0 \tag{11-51}$$

$$CsE + \frac{dI}{dx} = 0 \tag{11-52}$$

These can be combined to give

$$\frac{d^2E}{dx^2} - LCs^2E = 0 \tag{11-53}$$

Now let $\mathcal{L}_x[E(x,s)] \triangleq \bar{E}(\gamma,s)$, and (11-53) becomes

$$\gamma^2\bar{E} - LCs^2\bar{E} = \frac{dE(x,s)}{dx}\bigg|_{x=0} + \gamma E(0,s) \tag{11-54}$$

We can eliminate the derivative by using (11-51), so that (11-54) becomes

$$\bar{E}(\gamma,s) = \frac{\gamma E(0,s) - LsI(0,s)}{\gamma^2 - LCs^2} \tag{11-55}$$

We now take $\mathcal{L}_\gamma^{-1}$ using the tables in Appendix B,

$$\mathcal{L}_\gamma^{-1}[\bar{E}(\gamma,s)] = E(x,s) = E(0,s) \cosh \sqrt{LC}\, sx - \sqrt{\frac{L}{C}}\, I(0,s) \sinh \sqrt{LC}\, sx \tag{11-56}$$

The parameters in this equation are usually written

$$LC = \frac{1}{v^2} \tag{11-57}$$

where v is the velocity of propagation, and

$$\sqrt{\frac{L}{C}} = Z_0 \tag{11-58}$$

Z_0 is called the characteristic impedance. The next step in solution is to take $\mathcal{L}_s^{-1}$. This will be left for an exercise.

11-5. Summary and Table

In the few examples discussed above, it should be noted that the $\mathcal{L}$ method of solution has the property of reducing the partial differential equation to an ordinary one. There are other methods of solution, and many applications not mentioned here. These are treated in texts on differential equations.[1]

TABLE 11-1. TRANSFORMS*

1. $\dfrac{1}{s^4 - a^4}$	$\dfrac{1}{2a^3}$ (sinh at $-$ sin at)
2. $\dfrac{s}{s^4 - a^4}$	$\dfrac{1}{2a^2}$ (cosh at $-$ cos at)
3. $\dfrac{s^2}{s^4 - a^4}$	$\dfrac{1}{2a}$ (sinh at $+$ sin at)
4. $\dfrac{s^3}{s^4 - a^4}$	$\dfrac{1}{2}$ (cosh at $+$ cos at)
5. $\dfrac{2}{a} e^{-a\sqrt{s}}$	$\dfrac{e^{-a^2/4t}}{\sqrt{\pi t^3}}$
6. $\dfrac{e^{-a\sqrt{s}}}{\sqrt{s}}$	$\dfrac{e^{-a^2/4t}}{\sqrt{\pi t}}$
7. $\dfrac{e^{-a\sqrt{s}}}{s}$	$1 - \text{erf} \dfrac{a}{2\sqrt{t}} = \text{erfc} \dfrac{a}{2\sqrt{t}}$
8. $\dfrac{a}{s\sqrt{s} + a}$	$\dfrac{1}{\sqrt{a}} \text{erf} (\sqrt{at})$
9. $\dfrac{1}{\sqrt{s}\,(s - a)}$	$\dfrac{1}{\sqrt{a}} e^{at} \text{erf} (\sqrt{at})$
10. $\dfrac{1}{\sqrt{s^2 + a^2}\,(\sqrt{s^2 + a^2} + s)^n}$	$\dfrac{1}{a^n} J_n(at)$ (Bessel function)

* For an extensive table of irrational transforms, see A. Erdélyi, "Tables of Integral Transforms," McGraw-Hill Book Company, Inc., New York, 1954.

PROBLEMS

11-1. Write boundary conditions for a string of length L initially at rest with deflection shown:

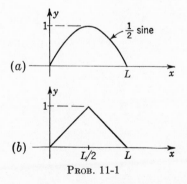

PROB. 11-1

[1] See, for example, K. S. Miller, "Partial Differential Equations in Engineering Problems," Prentice-Hall, Inc., Englewood Cliffs, N.J., 1953.

11-2. Solve for $y(x,t)$ by transforming twice:

$$\frac{\partial y}{\partial x} + \frac{\partial y}{\partial t} = 1 \qquad y(0,t) = y(x,0) = 0 \qquad x, t \geq 0$$

$$Ans. \quad y = \begin{cases} t, \ x > t \\ x, \ x < t \end{cases}$$

11-3. Solve for $y(x,t)$ by transforming twice:

$$\frac{\partial y}{\partial x} - \frac{\partial y}{\partial t} = 0 \qquad y(0,t) = t \qquad y(x,0) = x \qquad x, t \geq 0$$

$$Ans. \quad y = t + x$$

11-4. Sketch a three-dimensional plot of the solutions of Probs. 11-2 and 11-3 (see Fig. 11-2).

11-5. Write boundary conditions for heat conduction in a semi-infinite body if the temperature has an initial distribution $f(x)$ at time zero.   Assume that a step function of temperature is applied to the surface at time zero.

11-6. Transform the equation for heat conduction with conditions of Prob. 11-5 first with respect to x and then t.

11-7. Verify Eqs. (11-46) and (11-47).

11-8. Find $e(x,t)$ from (11-56). (*Hint*: Write the hyperbolic functions in exponential form.)

11-9. Find $E(x,s)$ as in (11-56) with the far end of the line open-circuited instead of shorted.

CHAPTER 12

FOURIER SERIES

In this chapter we shall discuss the expansion of a periodic function in a Fourier series. This subject is a familiar one, but we shall look at Fourier analysis as another kind of transform—the *finite Fourier transform.*

12-1. Fourier Series

If $f(t)$ is a periodic function of period T, then $f(t) = f(t + T)$. Such a function can be represented by a sum of sines and cosines:

$$f(t) = \frac{a_0}{T} + \frac{2}{T} \sum_{n=1}^{\infty} \left(a_n \cos \frac{2\pi n}{T} t + b_n \sin \frac{2\pi n}{T} t \right) \qquad (12\text{-}1)$$

where
$$a_n = \int_{-T/2}^{T/2} f(t) \cos \frac{2\pi n}{T} t \, dt$$
$$b_n = \int_{-T/2}^{T/2} f(t) \sin \frac{2\pi n}{T} t \, dt \qquad (12\text{-}2)$$

This expansion is valid if $f(t)$ has a finite number of maxima and minima in a period and if it has also a finite number of finite discontinuities. These requirements are called *Dirichlet's conditions.*[1]

The complex form of the Fourier expansion is somewhat simpler:

$$f(t) = \frac{1}{T} \sum_{n=-\infty}^{\infty} c_n e^{j\omega_n t} \qquad (12\text{-}3)$$

$$c_n = \int_{-T/2}^{T/2} f(t) e^{-j\omega_n t} \, dt \qquad \omega_n \triangleq \frac{2\pi n}{T} \qquad (12\text{-}4)$$

Since $f(t)$ is periodic, as is exp $(-j\omega_n t)$, we can write (12-4) as

$$c_n = \int_0^T f(t) e^{-j\omega_n t} \, dt \qquad (12\text{-}5)$$

---

[1] See, for example, I. N. Sneddon, "Fourier Transforms," p. 9, McGraw-Hill Book Company, Inc., New York, 1951.

which is in convenient form for comparison with the $\mathcal{L}$ formula. If we define a function

$$f_T(t) \triangleq \begin{cases} f(t) & 0 \le t \le T \\ 0 & \text{elsewhere} \end{cases} \tag{12-6}$$

then we can write

$$c_n = \mathcal{L}[f_T(t)]_{s=j\omega_n} \tag{12-7}$$

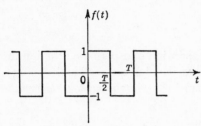

FIG. 12-1. Square wave.

Example 1. As an example, let us find the Fourier series expansion for the square wave shown in Fig. 12-1. We can write $f_T(t)$ in terms of step functions:

$$f_T(t) = u(t) - 2u\left(t - \frac{T}{2}\right) + u(t - T) \tag{12-8}$$

so that from (12-7)

$$c_n = \left[\frac{1}{s} - \frac{2e^{-(T/2)s}}{s} + \frac{e^{-Ts}}{s}\right]_{s=j\omega_n}$$

$$= \frac{(1 - e^{-j\omega_n T/2})^2}{j\omega_n} = \frac{T(1 - e^{-j\pi n})^2}{j2\pi n} \tag{12-9}$$

We have finally

$$c_n = \begin{cases} 2T/j\pi n & n \text{ odd} \\ 0 & n \text{ even} \end{cases} \tag{12-10}$$

We now write from (12-3)

$$f(t) = \frac{1}{T} \sum_{n=-\infty}^{\infty} c_n e^{j\omega_n t}$$

$$= \frac{2}{j\pi}\left[\cdots - \frac{e^{-j\omega_3 t}}{3} - \frac{e^{-j\omega_1 t}}{1} + \frac{e^{j\omega_1 t}}{1} + \frac{e^{j\omega_3 t}}{3} + \cdots\right]$$

$$= \frac{4}{\pi}\left[\sin \omega_1 t + \frac{\sin \omega_3 t}{3} + \cdots\right] \qquad \omega_n = \frac{2\pi n}{T} \tag{12-11}$$

The first two terms of the series and their sum are shown in Fig. 12-2.[1] The availability of $\mathcal{L}$ tables makes finding the Fourier coefficient c_n from (12-7) easier than carrying out the integration (12-4).

Example 2. Another example is the Fourier series for the output of a full-wave rectifier shown in Fig. 12-3. The arcs are half cycles of sine

[1] This series with $T = 2\pi$ was also discussed in Sec. 8-2.

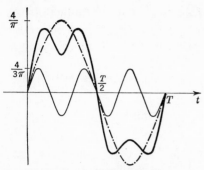

FIG. 12-2. First two Fourier series terms and their sum for square wave.

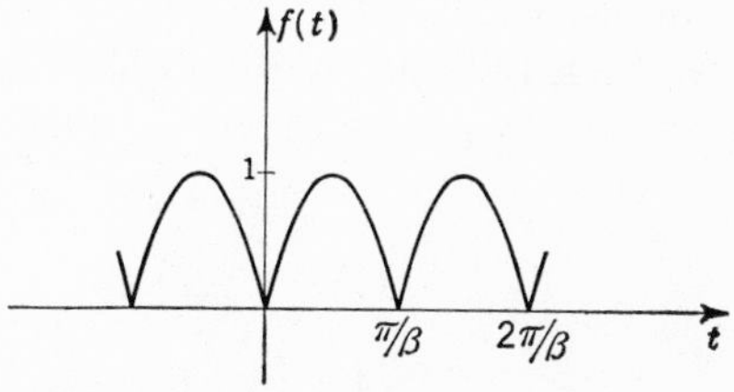

FIG. 12-3. Full-wave rectifier output.

waves.   In this case $T = \pi/\beta$, and we have

$$f_T(t) = \sin \beta t + u\left(t - \frac{\pi}{\beta}\right) \sin \beta \left(t - \frac{\pi}{\beta}\right) \tag{12-12}$$

so that, with $\omega_n = 2\pi n/T = 2\beta n$,

$$c_n = \left[\frac{\beta}{s^2 + \beta^2}\left(1 + e^{-(\pi/\beta)s}\right)\right]_{s=j\omega_n}$$

$$= \frac{\beta}{-4n^2\beta^2 + \beta^2}\left(1 + e^{-j2\pi n}\right) = \frac{2}{\beta(1 - 4n^2)} \tag{12-13}$$

Then

$$f(t) = \frac{1}{T}\sum_{n=-\infty}^{\infty} c_n e^{j\omega_n t}$$

$$= \cdots - \frac{2}{15\pi}e^{-j\omega_2 t} - \frac{2}{3\pi}e^{-j\omega_1 t} + \frac{2}{\pi} - \frac{2}{3\pi}e^{j\omega_1 t} - \frac{2}{15\pi}e^{j\omega_2 t} - \cdots$$

$$= \frac{2}{\pi}\left[1 - \frac{2}{3}\cos 2\beta t - \frac{2}{15}\cos 4\beta t - \cdots\right] \tag{12-14}$$

12-2. Finite Fourier Transforms

The formulas (12-3) and (12-4) can be looked upon as a transform of $f(t)$ and its inverse.   Let us write

$$\mathbf{F}[f(t)] \triangleq \int_{-T/2}^{T/2} f(t)e^{-j\omega_n t}\,dt \triangleq \hat{f}(n) \tag{12-15}$$

$$\mathbf{F}^{-1}[\hat{f}(n)] = \frac{1}{T}\sum_{n=-\infty}^{\infty} \hat{f}(n)e^{j\omega_n t} = f(t) \tag{12-16}$$

$$\omega_n \triangleq \frac{2\pi n}{T}$$

We shall call $\mathbf{F}[f(t)]$ the *finite Fourier transform* of $f(t)$. As in the case of the $\mathcal{L}$ transform, the transform $\hat{f}(n)$ is obtained from $f(t)$ as the result of a mathematical operation. The transform variable in this case is n, an integer.

We can put the results of the examples of the preceding section into transform-pair form (Table 12-1).

TABLE 12-1. F-TRANSFORM PAIRS

$f(t) \quad 0 \le t \le T$	$\hat{f}(n)$
$\begin{cases} 1 & 0 < t < \dfrac{T}{2} \\ -1 & \dfrac{T}{2} < t < T \end{cases}$	$\begin{cases} \dfrac{2T}{j\pi n} & n \text{ odd} \\ 0 & n \text{ even} \end{cases}$
$\sin \beta t \qquad \beta = \dfrac{\pi}{T}$	$\dfrac{2}{\beta(1 - 4n^2)}$

So far, writing the Fourier series as a transform is just another way of setting down a familiar relationship. The transform does have a purpose, however, since it can be used to solve differential equations in the same way as $\mathcal{L}$ transforms. We shall need the $\mathbf{F}$ transform of derivatives of $f(t)$ for this.

We write from (12-15)

$$\mathbf{F}[f'(t)] = \int_{-T/2}^{T/2} f'(t)e^{-j\omega_n t}\, dt \tag{12-17}$$

Integrating by parts, we have

$$\mathbf{F}[f'(t)] = f(t)e^{-j\omega_n t} \Big|_{-T/2}^{T/2} + j\omega_n \int_{-T/2}^{T/2} f(t)e^{-j\omega_n t}\, dt$$

$$= \left[f\left(\frac{T}{2}\right)e^{-j\pi n} - f\left(-\frac{T}{2}\right)e^{j\pi n} \right] + j\omega_n \hat{f}(n)$$

$$= j\omega_n \hat{f}(n) \tag{12-18}$$

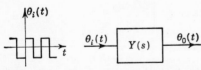

FIG. 12-4. System subjected to a periodic input.

since $f(T/2) = f(-T/2)$. We see that the derivative of $f(t)$ has a transform, but unlike $\mathcal{L}$ transforms, the initial conditions do not appear. The transform of the kth derivative can be found in the same way.

$$\mathbf{F}[f^{(k)}(t)] = (j\omega_n)^k \hat{f}(n) \qquad \omega_n = \frac{2\pi n}{T} \tag{12-19}$$

Let us now consider the system shown in Fig. 12-4.

From the definition of the transfer function $Y(s)$ we can write, assuming for illustration a simple rational fraction for $Y(s)$:

$$\frac{\mathcal{L}[\theta_0(t)]}{\mathcal{L}[\theta_i(t)]} \triangleq Y(s) = \frac{A_1 s + A_0}{B_2 s^2 + B_1 s + B_0} \tag{12-20}$$

If we assume that all initial conditions are zero, we can multiply out (12-20) to get the differential equation

$$B_2\theta_0''(t) + B_1\theta_0'(t) + B_0\theta_0(t) = A_1\theta_i'(t) + A_0\theta_i(t) \qquad (12\text{-}21)$$

Now if we assume that both θ_0 and θ_i are periodic—which is to say all the transients have died out and the system is in the steady state—we can take the **F** transform of both sides of (12-21) and obtain

$$[B_2(j\omega_n)^2 + B_1j\omega_n + B_0]\hat{\theta}_0(n) = (A_1j\omega_n + A_0)\hat{\theta}_i(n) \qquad (12\text{-}22)$$

The new transfer function (the ratio of **F** transforms of output and input) for periodic inputs in the steady state can be written down:

$$\frac{\mathbf{F}[\theta_0(t)]}{\mathbf{F}[\theta_i(t)]} = \frac{A_1j\omega_n + A_0}{B_2(j\omega_n)^2 + B_1j\omega_n + B_0} = Y(j\omega_n) \qquad (12\text{-}23)$$

If the square wave of Fig. 12-1 is applied to this system, then from Table 12-1

$$\mathbf{F}[\theta_0(t)] = \begin{cases} \dfrac{2T}{j\pi n} \dfrac{A_1j(2\pi n/T) + A_0}{B_2[j(2\pi n/T)]^2 + B_1j(2\pi n/T) + B_0} & n \text{ odd} \\ 0 & n \text{ even} \end{cases} \qquad (12\text{-}24)$$

The output $\theta_0(t)$ is the Fourier series found from (12-16). The first terms of the series in this case are (for $n = \pm1$)

$$\theta_{0_1}(t) = \frac{1}{T}\left[-\frac{2T}{j\pi} Y\left(-\frac{2\pi j}{T}\right) e^{-j(2\pi/T)t} + \frac{2T}{j\pi} Y\left(\frac{2\pi j}{T}\right) e^{j(2\pi/T)t} \right] \qquad (12\text{-}25)$$

Since the right side of (12-25) is the sum of conjugate terms, we can write

$$\begin{aligned} \theta_{0_1}(t) &= \frac{4}{\pi} \operatorname{Re}\left[\frac{e^{j(2\pi/T)t}}{j} Y\left(\frac{2\pi j}{T}\right) \right] \\ &= \frac{4}{\pi}\left| Y\left(\frac{2\pi j}{T}\right) \right| \sin\left(\frac{2\pi}{T} t + \phi\right) \\ \phi &= \angle\left[Y\left(\frac{2\pi j}{T}\right) \right] \end{aligned} \qquad (12\text{-}26)$$

The other terms can be found in the same way.

12-3. $\mathbf{F}_s$ and $\mathbf{F}_c$ Transforms

Let us discuss the application of the **F**-transform method to the solution of partial differential equations. We shall need to modify the **F** transform to bring in initial conditions, but first let us look at a typical problem.

An infinite slab of thickness L has an initial temperature distribution $f(x)$. The ends of the slab are kept at zero temperature. The equation for the problem is, from Sec. 11-3,

$$\frac{\partial^2\theta}{\partial x^2} - \frac{1}{a}\frac{\partial\theta}{\partial t} = 0$$

$$\theta(0,t) = \theta(L,t) = 0$$

$$\theta(x,0) = f(x)$$

(12-27)

where $\theta(x,t)$ is the temperature and a is thermal diffusivity. The initial temperature of the slab might be as shown in Fig. 12-5.

In this problem time ranges from zero to infinity, whereas we are interested in values of x between 0 and L. Therefore, while $\mathcal{L}$ transform seems appropriate for the *time* part of the problem, the *space* part might be better handled in another way. The fact that the initial temperature

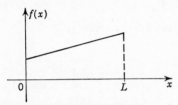

FIG. 12-5. Initial temperature of the slab.

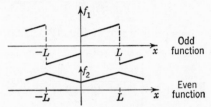

FIG. 12-6. Two ways of making $f(x)$ periodic. The period is $2L$.

is not periodic in x will not keep us from using $\mathbf{F}$ transforms. We can assume it periodic, making the nature of the periodicity consistent with the problem. Two ways of making $f(x)$ periodic are shown in Fig. 12-6. As we shall see, the choice of which to use will be taken care of by the transform.

If we make use of the fact that in the figure f_1 is odd and f_2 is even, we can simplify the Fourier expansion given by (12-1) and (12-2).

For odd functions we can write $a_n = 0$, so that (12-1) and (12-2) take the form of a transform pair:

$$f_1(x) = \frac{2}{L}\sum_{n=1}^{\infty} \hat{f}_s(n) \sin \omega_n x \qquad \omega_n = \frac{n\pi}{L}$$

$$\mathbf{F}_s[f_1(x)] \triangleq \hat{f}_s(n) = \int_0^L f(x) \sin \omega_n x \, dx$$

(12-28)

For even functions, $b_n = 0$ and we have

$$f_2(x) = \frac{\hat{f}_c(0)}{L} + \frac{2}{L}\sum_{n=1}^{\infty} \hat{f}_c(n) \cos \omega_n x \qquad \omega_n = \frac{n\pi}{L}$$

$$\mathbf{F}_c[f_2(x)] \triangleq \hat{f}_c(n) = \int_0^L f(x) \cos \omega_n x \, dx$$

(12-29)

We have written $f(x)$ in each of the integrands because over the range of integration the functions of Fig. 12-6 are equal. Formulas (12-28) and (12-29) are called the *finite sine* and *finite cosine transforms*.[1] Their use in a specific problem is governed by initial conditions, which will now appear in the transforms of derivatives.

For the F_s and F_c transforms we have, integrating by parts,

$$\mathbf{F}_s[y'(x)] = \int_0^L y'(x) \sin \omega_n x \, dx$$

$$= [y(x) \sin \omega_n x]_0^L - \omega_n \int_0^L y(x) \cos \omega_n x \, dx$$

$$= -\omega_n \mathbf{F}_c[y(x)] \tag{12-30}$$

and

$$\mathbf{F}_c[y'(x)] = \int_0^L y'(x) \cos \omega_n x \, dx$$

$$= [y(x) \cos \omega_n x]_0^L + \omega_n \int_0^L y(x) \sin \omega_n x \, dx$$

$$= [(-)^n y(L) - y(0)] + \omega_n \mathbf{F}_s[y(x)] \tag{12-31}$$

Using these results, we can find the transforms of higher derivatives:

$$\mathbf{F}_s[y''(x)] = -\omega_n \mathbf{F}_c[y'(x)]$$

$$= -\omega_n^2 \mathbf{F}_s[y(x)] - \omega_n[(-)^n y(L) - y(0)] \tag{12-32}$$

and

$$\mathbf{F}_c[y''(x)] = \omega_n \mathbf{F}_s[y'(x)] + [(-)^n y'(L) - y'(0)]$$

$$= -\omega_n^2 \mathbf{F}_c[y(x)] + [(-)^n y'(L) - y'(0)] \tag{12-33}$$

Evidently in second-order equations when the value of y is given at 0 and L, the F_s transform should be used. When the value of y' is given at 0 and L, we use the F_c transform.

It is interesting to notice that the F_c and F_s transforms allow us to introduce boundary conditions at two points 0 and L as compared with $\mathcal{L}$ transforms which take care of one-point boundary conditions.

The F_s and F_c transforms of a function can be found from the $\mathcal{L}$ transform in the following way. We introduce $f_L(x)$:

$$f_L(x) \triangleq \begin{cases} f(x) & 0 < x < L \\ 0 & \text{elsewhere} \end{cases} \tag{12-34}$$

Then we can write from (12-28) and (12-29)

$$\mathcal{L}[f_L(x)]_{s=j\omega_n} = \int_0^L f(x) e^{-j\omega_n x} \, dx$$

$$= \int_0^L f(x) \cos \omega_n x \, dx - j \int_0^L f(x) \sin \omega_n x \, dx$$

$$= \mathbf{F}_c[f(x)] - j\mathbf{F}_s[f(x)] \tag{12-35}$$

---

[1] See Sneddon, *op. cit.*, for many applications of these transforms.

So that we have the relations, with $\omega_n \triangleq n\pi/L$,

$$\mathbf{F}_c[f(x)] = \text{Re } \{\mathcal{L}[f_L(x)]_{s=j\omega_n}\} \tag{12-36}$$

and
$$\mathbf{F}_s[f(x)] = -\text{Im } \{\mathcal{L}[f_L(x)]_{s=j\omega_n}\} \tag{12-37}$$

12-4. Solution of a Heat-conduction Problem

We return now to the problem described in (12-27). Let us first take the $\mathcal{L}$ transform to make this an ordinary differential equation. We shall let $\mathcal{L}[\theta(x,t)] \triangleq \theta_1(x,s)$. The transformed equation is

$$\frac{d^2\theta_1}{dx^2} - \frac{s}{a}\,\theta_1 = -\frac{1}{a}\,\theta(x,0) = -\frac{1}{a}\,f(x) \tag{12-38}$$

Now the boundary conditions give the value of θ at 0 and L, so that the $\mathbf{F}_s$ transform should be used. We let $\theta_2(n,s) \triangleq \mathbf{F}_s[\theta_1(x,s)]$ and transform (12-38):

$$-\omega_n{}^2\theta_2 - \frac{s}{a}\,\theta_2 = -\frac{1}{a}\,\mathbf{F}_s[f(x)] + \omega_n[(-)^n\theta_1(L,s) - \theta_1(0,s)] \tag{12-39}$$

The term in the brackets is zero from the statement of the problem (12-27). Solving (12-39) we have

$$\theta_2(n,s) = \frac{(1/a)\mathbf{F}_s[f(x)]}{\omega_n{}^2 + (s/a)} \tag{12-40}$$

We shall take $\mathcal{L}^{-1}$ of (12-40) first, letting $\mathcal{L}^{-1}[\theta_2(n,s)] \triangleq \theta_3(n,t)$:

$$\theta_3(n,t) = \mathbf{F}_s[f(x)]e^{-a\omega_n{}^2t} \tag{12-41}$$

The final inversion will be performed using (12-28). First let us assume a form for $f(x)$. Let

$$f(x) = kx \qquad 0 < x < L \tag{12-42}$$

Then from (12-37)

$$\mathbf{F}_s[f(x)] = -\text{Im}\left[\frac{k}{s^2}(1 - e^{-sL}) - kL\frac{e^{-sL}}{s}\right]_{s=j\omega_n} \qquad \omega_n = \frac{n\pi}{L}$$

$$= -k\,\text{Im}\left[\frac{1 - \cos\omega_nL + j\sin\omega_nL}{-\omega_n{}^2} - L\frac{\cos\omega_nL - j\sin\omega_nL}{j\omega_n}\right]$$

$$= \frac{k\sin\omega_nL}{\omega_n{}^2} - kL\frac{\cos\omega_nL}{\omega_n}$$

$$= \frac{k\sin n\pi}{\omega_n{}^2} - kL\frac{\cos n\pi}{\omega_n} = \frac{kL(-)^{n+1}}{\omega_n} \tag{12-43}$$

Our result (12-41) now becomes

$$\theta_3(n,t) = \frac{(-)^{n+1}kLe^{-a\omega_n{}^2t}}{\omega_n} \tag{12-44}$$

and from (12-28) we have finally

$$\theta(x,t) = \frac{2kL}{L} \sum_{n=1}^{\infty} \frac{(-)^{n+1}e^{-a\omega_n^2 t}}{\omega_n} \sin \omega_n x$$

$$= \frac{2kL}{\pi} \sum_{n=1}^{\infty} \frac{(-)^{n+1}}{n} e^{-a(n^2\pi^2/L^2)t} \sin \frac{n\pi}{L} x \qquad (12\text{-}45)$$

12-5. The Fourier Integral

We can change a periodic function into a nonperiodic one by letting the period become large without bound. This process is illustrated for a train of rectangular pulses in Fig. 12-7. We shall now consider the effect of this expansion process on the representation of $f(t)$ by the Fourier series (12-3):

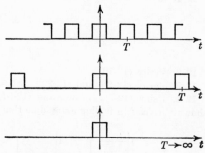

FIG. 12-7. Effect of increasing the period of a pulse train.

$$f(t) = \frac{1}{T} \sum_{n=-\infty}^{\infty} c_n e^{j\omega_n t} \qquad \omega_n = \frac{2n\pi}{T}$$

$$c_n = \int_{-T/2}^{T/2} f(t)e^{-j\omega_n t} \, dt \qquad (12\text{-}46)$$

We note that an increment in ω_n, $\Delta\omega_n$, is

$$\Delta\omega_n = \frac{2(n+1)\pi}{T} - \frac{2n\pi}{T} = \frac{2\pi}{T} \triangleq 2\pi \, \Delta f_n \qquad (12\text{-}47)$$

Now we write (12-46)

$$f(t) = \sum_{-\infty}^{\infty} c_n e^{j\omega_n t} \, \Delta f_n$$

$$c_n = \int_{-T/2}^{T/2} f(t)e^{-j\omega_n t} \, dt \qquad (12\text{-}48)$$

If we now let $T \to \infty$ in such a way that ω_n approaches a continuous variable ω, the sum becomes formally an integral and we write

$$f(t) = \int_{-\infty}^{\infty} F(\omega)e^{j\omega t} \, df \qquad \omega \triangleq 2\pi f$$

$$F(\omega) = \int_{-\infty}^{\infty} f(t)e^{-j\omega t} \, dt \qquad (12\text{-}49)$$

Equations (12-49) define the *Fourier transform* and represent the expansion of a function, not necessarily periodic, in terms of sines and cosines.

Instead of sinusoids of discrete frequencies adding together to become $f(t)$ as in the Fourier series expansion, we have all frequencies present. The amplitude of a constituent of $f(t)$ at frequency f_1 is $F(\omega_1)\,df$.

In the next chapter we shall discuss the properties of the Fourier transform and some of its applications.

PROBLEMS

12-1. Show from (12-2) and (12-4) that

$$a_n = \frac{c_n + c_{-n}}{2}$$

$$-jb_n = \frac{(c_n - c_{-n})}{2}$$

12-2. Verify (12-10).

12-3. Verify (12-28) and (12-29). Show that $f_s = b_n/2$ in (12-1) and (12-2).

12-4. Using (12-7), (12-36), and (12-37) find c_n, $\hat{f}_c(n)$, and $\hat{f}_s(n)$ for the functions shown. (*Note:* In finding c_n, assume that L is a period.)

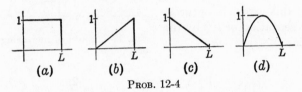

PROB. 12-4

12-5. Assume that a string is stretched initially as shown. It is fixed at 0 and L, is initially at rest, and then is released at $t = 0$. Find $y(x,t)$, using finite transforms.

$$Ans. \quad y(x,t) = \left(\frac{8}{\pi^2}\right)\left(\cos\frac{\pi ct}{L}\sin\frac{\pi x}{L} - \frac{1}{9}\cos\frac{3\pi ct}{L}\sin\frac{3\pi x}{L} + \cdots\right)$$

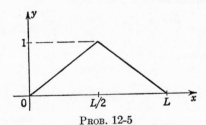

PROB. 12-5

12-6. Find the magnitude of the first three harmonics of the output of the filter shown when the square wave of Fig. 12-1 is applied.

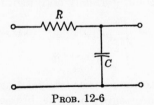

PROB. 12-6

12-7.[1] Assume that a periodic function $f(t)$ is to be approximated by a finite sum $f_N(t)$ where

$$f_N(t) = \frac{1}{T} \sum_{n=0}^{N} A_n \exp(\omega_n t) \qquad \omega_n = \frac{2\pi n}{T}$$

Find values of A_n that make the mean square error $\dfrac{1}{T} \displaystyle\int_0^T [f(t) - f_N(t)]^2 \, dt$ a minimum. Show that these are the Fourier coefficients c_n.

[1] See R. V. Churchill, "Fourier Series and Boundary Value Problems," pp. 40-42, McGraw-Hill Book Company, Inc., New York, 1941.

CHAPTER 13

FOURIER TRANSFORMS

We saw in Sec. 12-5 how the formulas for Fourier series expansion of a periodic function change when the period becomes infinite. The Fourier integral formulas which resulted form a transform pair which we call the *Fourier transform;*

$$\mathfrak{F}[f(t)] \triangleq \int_{-\infty}^{\infty} f(t)e^{-j\omega t} \, dt \triangleq F(\omega) \tag{13-1}$$

$$\mathfrak{F}^{-1}[F(\omega)] = \int_{-\infty}^{\infty} F(\omega)e^{j\omega t} \, df = f(t) \qquad \omega \triangleq 2\pi f \tag{13-2}$$

In this chapter we shall discuss some of the properties and applications of the Fourier transform.

13-1. Fourier Transforms of Functions

We can proceed to develop the theory of Fourier transforms just as we did for Laplace transforms. The formula (13-1) defines $\mathfrak{F}[f(t)]$ as the result of a mathematical operation on $f(t)$ in the same way as $\mathcal{L}[f(t)]$ is defined for another operation. The condition for existence of $F(\omega)$ is usually given as

$$\int_{-\infty}^{\infty} |f(t)| \, dt < \infty \tag{13-3}$$

Functions which do not satisfy (13-3) may have Fourier transforms, however. For example, functions for which the average power

$$\lim_{T \to \infty} \frac{1}{2T} \int_{-T}^{T} f^2(t) \, dt \tag{13-4}$$

is bounded and which have statistical properties which do not change with time can have $\mathfrak{F}$ transforms containing impulse functions. The sinusoid is an example of a function which fails to meet (13-3) but does have a Fourier transform. We shall discuss functions like this in detail in Chap. 15.

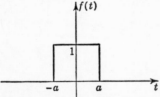

FIG. 13-1. Pulse to be transformed.

Example 1. The Fourier transform of a function can be illustrated by the following example. Let us transform $f(t)$ shown in Fig. 13-1. From

(13-1) we have

$$\mathcal{F}[f(t)] = \int_{-a}^{a} e^{-i\omega t}\, dt = \int_{0}^{a} e^{i\omega t}\, dt + \int_{0}^{a} e^{-i\omega t}\, dt$$

$$= 2 \int_{0}^{a} \cos \omega t\, dt = \frac{2 \sin \omega a}{\omega} \tag{13-5}$$

Example 2: Impulse Functions. The transforms of impulse functions lead to some useful results:

$$\mathcal{F}[\delta(t - t_1)] = \int_{-\infty}^{\infty} \delta(t - t_1) e^{-i\omega t}\, dt = e^{-i\omega t_1} \tag{13-6}$$

We can now take the inverse Fourier transform of both sides of (13-6):

$$\mathcal{F}^{-1}\{\mathcal{F}[\delta(t - t_1)]\} = \delta(t - t_1) = \int_{-\infty}^{\infty} e^{-i\omega t_1} e^{i\omega t}\, df$$

$$= \int_{-\infty}^{\infty} e^{i\omega(t - t_1)}\, df \tag{13-7}$$

This expression[1] for the impulse function in time and the corresponding one for frequency are written below for reference:

$$\delta(t - t_1) = \int_{-\infty}^{\infty} e^{i\omega(t - t_1)}\, df = \mathcal{F}^{-1}[e^{-i\omega t_1}] \tag{13-8}$$

$$\delta(f - f_1) = \int_{-\infty}^{\infty} e^{-it(\omega - \omega_1)}\, dt = \mathcal{F}[e^{i\omega_1 t}] \tag{13-9}$$

Since $\delta(x) = \delta(-x)$, we can modify these formulas when necessary. Another form which is sometimes useful is

$$\delta(f - f_1) + \delta(f + f_1) = 2 \int_{-\infty}^{\infty} \cos \omega t \cos \omega_1 t\, dt$$

$$= 4 \int_{0}^{\infty} \cos \omega t \cos \omega_1 t\, dt \tag{13-10}$$

When $f_1 = 0$,

$$\delta(f) = \int_{-\infty}^{\infty} \cos \omega t\, dt = 2 \int_{0}^{\infty} \cos \omega t\, dt \tag{13-11}$$

Example 3. Now let us transform a constant K:

$$\mathcal{F}[K] = K \int_{-\infty}^{\infty} e^{-i\omega t}\, dt = K\delta(f) \tag{13-12}$$

where we have used (13-9) in obtaining the result. This is an example of an $f(t)$ which satisfies (13-4) but not (13-3).

---

[1] For a rigorous discussion of integrals of this type, see B. Friedman, "Principles and Techniques of Applied Mathematics," p. 220, John Wiley & Sons, Inc., NewYork, 1956.

13-2. Solution of Differential Equations

The Fourier transform of a derivative can be found formally as follows. We write from (13-2)

$$f(t) = \int_{-\infty}^{\infty} F(\omega)e^{j\omega t}\, df \qquad (13\text{-}13)$$

and simply differentiate both sides with respect to t:

$$f'(t) = \int_{-\infty}^{\infty} j\omega F(\omega)e^{j\omega t}\, df \qquad (13\text{-}14)$$

Evidently, then,[1]

$$\mathfrak{F}\left[\frac{df(t)}{dt}\right] = j\omega \mathfrak{F}[f(t)] \qquad (13\text{-}15)$$

This result can be generalized to

$$\mathfrak{F}\left[\frac{d^n f(t)}{dt^n}\right] = (j\omega)^n \mathfrak{F}[f(t)] \qquad (13\text{-}16)$$

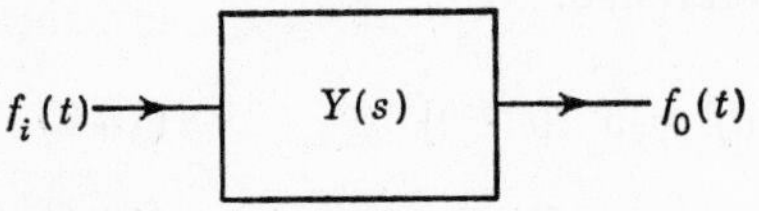

Now let us consider the system shown in Fig. 13-2. If, for example,

$f_i(t) \rightarrow$ [$Y(s)$] $\rightarrow f_0(t)$

FIG. 13-2. Linear-system block diagram.

$$Y(s) = \frac{A_1 s + A_0}{B_2 s^2 + B_1 s + B_0} \qquad (13\text{-}17)$$

the differential equation for the system is

$$B_2 f_0''(t) + B_1 f_0'(t) + B_0 f_0(t) = A_1 f_i'(t) + A_0 f_i(t) \qquad (13\text{-}18)$$

Now from (13-16) the Fourier transform of the differential equation is

$$(j\omega)^2 B_2 F_0(\omega) + j\omega B_1 F_0(\omega) + B_0 F_0(\omega) = A_1 j\omega F_i(\omega) + A_0 F_i(\omega) \qquad (13\text{-}19)$$

where $F_i(\omega) = \mathfrak{F}[f_i(t)]$, etc. The transfer function from the Fourier-transformed equation is

$$\frac{\mathfrak{F}\,[\text{output}]}{\mathfrak{F}\,[\text{input}]} = \frac{A_1 j\omega + A_0}{B_2(j\omega)^2 + B_1 j\omega + B_0} = Y(j\omega) \qquad (13\text{-}20)$$

We can generalize this result to say

The Fourier transfer function is the Laplace transfer function with $s = j\omega$.

The Fourier transform can be used in the same way as the $\mathcal{L}$ transform to solve differential equations. The chief differences lie in the restricted class of functions which can be $\mathfrak{F}$ transformed and in the way the initial conditions must be introduced.

[1] This result holds for $f(t)$s satisfying (13-4) as well as (13-3).

Example.   As an example, let us consider the case where

$$Y(s) = \frac{1}{s + a} \tag{13-21}$$

and
$$f_i(t) = \begin{cases} e^{-\alpha t} & t \geq 0 \\ 0 & t < 0 \end{cases}$$
$$f_0(0) = b \tag{13-22}$$

Lumping the initial condition with the input according to Sec. 3-7, we have a new input

$$f_{i_1} = b\delta_+(t) + e^{-\alpha t}u(t) \tag{13-23}$$

Using the $\mathcal{L}$ transform, the output is given by

$$f_0 = \mathcal{L}^{-1}\left[\frac{b}{s + a} + \frac{1}{(s + \alpha)(s + a)} \right]$$
$$= be^{-at} + \frac{1}{a - \alpha}(e^{-\alpha t} - e^{-at}) \tag{13-24}$$

Now the same problem can be solved by Fourier transforms.   Writing the differential equation,

$$f_0' + af_0 = f_i = b\delta_+(t) + e^{-\alpha t}u(t) \tag{13-25}$$

The Fourier transform is

$$(j\omega + a)F_0(\omega) = b + \int_0^\infty e^{-\alpha t}e^{-j\omega t}\, dt$$
$$= b + \frac{1}{\alpha + j\omega} \tag{13-26}$$

Then
$$F_0(\omega) = \frac{b}{j\omega + a} + \frac{1}{(j\omega + a)(j\omega + \alpha)} \tag{13-27}$$

Using the $\mathcal{F}$ pair developed in (13-26),

$$e^{-\alpha t}u(t) \qquad \Bigg| \qquad \frac{1}{j\omega + \alpha} \tag{13-28}$$

the result of the inversion of (13-27) will be the same as (13-24).   We see that when the functions involved have Fourier transforms, the Fourier method can be used to solve differential equations.

In contrast to this example there is a type of problem which can be treated with Fourier transforms more easily than by Laplace.   This is the determination of the "steady-state" solution—the subject of the next section.

13-3. Steady State

In Sec. 9-4 we saw that the steady-state amplitude of the sinusoidal response of a system with transfer function $Y(s)$ was $|Y(j\omega)|$ and the phase angle $\angle[Y(j\omega)]$. Let us approach the problem from a different point of view. We consider the system shown in Fig. 13-3. The input is $e^{j\omega_1 t}$, so that $x_0(t)$ will contain the response to both sine and cosine. The output is given by

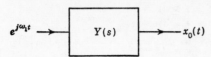

FIG. 13-3. System with sinusoidal input.

$$x_0(t) = \mathcal{L}^{-1}\left[Y(s) \cdot \frac{1}{s - j\omega_1} \right] \tag{13-29}$$

which becomes, using the convolution theorem (8-50),

$$x_0(t) = \int_0^t h(\tau)e^{j\omega_1(t-\tau)}\,d\tau \tag{13-30}$$

where

$$h(t) \triangleq \mathcal{L}^{-1}[Y(s)] \tag{13-31}$$

We write (13-30)

$$\begin{aligned}
x_0(t) &= e^{j\omega_1 t}\int_0^t h(\tau)e^{-j\omega_1\tau}\,d\tau \\
&= e^{j\omega_1 t}\left[\int_0^\infty h(\tau)e^{-j\omega_1\tau}\,d\tau - \int_t^\infty h(\tau)e^{-j\omega_1\tau}\,d\tau \right]
\end{aligned} \tag{13-32}$$

Now if the impulse response $h(t)$ dies out with time, the first integral of (13-32) exists and is independent of time. The second integral is the transient part of the solution. We now write

$$x_0(t) = e^{j\omega_1 t}\left[Y(j\omega_1) - \int_t^\infty h(\tau)e^{-j\omega_1\tau}\,d\tau \right] \tag{13-33}$$

and identify the term

$$x_0(t)_{ss} \triangleq e^{j\omega_1 t}Y(j\omega_1) \tag{13-34}$$

as the *steady-state* part of the output.

Suppose, now, we had applied the Fourier-transform method to the differential equation describing the system of Fig. 13-3. Since the integral is defined over all time, we can assume that the sinusoidal input has been applied continuously and that there are no initial conditions. We now write from (13-20)

$$\mathcal{F}[x_0(t)] = Y(j\omega)\mathcal{F}[e^{j\omega_1 t}] \tag{13-35}$$

But from (13-9),

$$\mathcal{F}[e^{j\omega_1 t}] = \delta(f - f_1) \tag{13-36}$$

so that

$$\mathcal{F}[x_0(t)] = Y(j\omega)\delta(f - f_1) \tag{13-37}$$

and

$$x_0(t) = \int_{-\infty}^\infty Y(j\omega)e^{j\omega t}\delta(f - f_1)\,df = e^{j\omega_1 t}Y(j\omega_1) \tag{13-38}$$

This result is the same as (13-34), so we can say:

When a periodic input is applied to a system, the Fourier-transform method will give the steady-state output.

The input need not be a single sinusoid but can be a combination. For example, if the input is

$$x_i(t) = \sum_n a_n e^{j\omega_n t} \tag{13-39}$$

then

$$x_0(t)_{ss} = \sum_n a_n Y(j\omega_n) e^{j\omega_n t} \tag{13-40}$$

13-4. Convolution and Other Properties

The Fourier transform of a product can be found as in the case of the Laplace transform. We write, with $\mathfrak{F}[f_1(t)] \triangleq F_1(\omega)$, etc.,

$$\mathfrak{F}[f_1(t)f_2(t)] = \int_{-\infty}^{\infty} dt\, e^{-j\omega t} \int_{-\infty}^{\infty} F_1(\omega_1)e^{j\omega_1 t}\, df_1 \int_{-\infty}^{\infty} F_2(\omega_2)e^{j\omega_2 t}\, df_2 \tag{13-41}$$

Upon interchange of order of integration we have

$$
\begin{aligned}
\mathfrak{F}[f_1(t)f_2(t)] &= \int_{-\infty}^{\infty} df_1 \int_{-\infty}^{\infty} df_2\, F_1(\omega_1)F_2(\omega_2) \int_{-\infty}^{\infty} e^{-j(\omega-\omega_1-\omega_2)t}\, dt \\
&= \int_{-\infty}^{\infty} df_1 \int_{-\infty}^{\infty} df_2\, F_1(\omega_1)F_2(\omega_2)\delta(f - f_1 - f_2) \\
&= \int_{-\infty}^{\infty} F_1(\omega_1)F_2(\omega - \omega_1)\, df_1 \\
&\qquad\qquad = \int_{-\infty}^{\infty} F_2(\omega_2)F_1(\omega - \omega_2)\, df_2 \tag{13-42}
\end{aligned}
$$

A similar expression can be derived for the inverse:

$$
\begin{aligned}
\mathfrak{F}^{-1}[F_1(\omega)F_2(\omega)] &= \int_{-\infty}^{\infty} f_1(t_1)f_2(t - t_1)\, dt_1 \\
&\qquad\qquad = \int_{-\infty}^{\infty} f_2(t_2)f_1(t - t_2)\, dt_2 \tag{13-43}
\end{aligned}
$$

Some of the other properties of the $\mathfrak{F}$ transform are tabulated in Table 13-1.

TABLE 13-1. PROPERTIES OF FOURIER TRANSFORMS

$f(t)$	$F(\omega)$
$f(-t)$	$F(-\omega)$
$f(at)$	$\dfrac{1}{a} F\left(\dfrac{\omega}{a}\right),\ a > 0$
$f(t)e^{-j\omega_1 t}$	$F(\omega + \omega_1)$
$f(t + t_1)$	$e^{j\omega t_1}F(\omega)$

13-5. Fourier and Laplace Transforms

A comparison of the formulas for Laplace and Fourier transforms shows considerable similarity. The formulas are

$$\mathcal{L}[f(t)] = \int_0^\infty f(t)e^{-st}\,dt \qquad s = \sigma + j\omega \tag{13-44}$$

and
$$\mathcal{F}[f(t)] = \int_{-\infty}^\infty f(t)e^{-j\omega t}\,dt \tag{13-45}$$

We make the following observations about the transforms:

1. For some $f(t)$s the formulas can be made the same. In particular, if

$$f(t) = 0 \qquad t < 0 \tag{13-46}$$

and
$$\int_0^\infty |f(t)|\,dt < \infty \tag{13-47}$$

so that σ in (13-44) can be set to zero, then

$$\mathcal{F}[f(t)] = \mathcal{L}[f(t)]_{s=j\omega} \tag{13-48}$$

2. Some $f(t)$s have $\mathcal{L}$ transforms but do not have $\mathcal{F}$ transforms. For example, the step function

$$f(t) = u(t) \tag{13-49}$$

has a $\mathcal{L}$ transform equal to $1/s$, but the integral

$$\int_0^\infty e^{-j\omega t}\,dt \tag{13-50}$$

does not exist, even if impulses are admitted. One might be tempted to replace $j\omega$ in (13-50) by $\sigma + j\omega$ and then let $\sigma \to 0$, in which case (13-50) becomes $1/j\omega$. This really amounts to taking the $\mathcal{L}$ transform of the step function, and the result cannot be called a Fourier transform.

3. An $f(t)$ which has a value for negative t cannot be uniquely represented as an inverse $\mathcal{L}$ transform. This is because we have defined $\mathcal{L}$ transforms for functions which are zero for $t < 0$. The Fourier transform, when it exists, repesents these negative time functions. Some authors use a form called *bilateral* $\mathcal{L}$ transforms to handle this situation.[1] If we write

$$\mathcal{L}_B[f(t)] \triangleq \int_{-\infty}^\infty e^{-st}f(t)\,dt \qquad s = \sigma + j\omega \tag{13-51}$$

and then choose appropriate values for σ when t is negative and positive, a transform results which will represent $f(t)$. Usually we write

$$\mathcal{L}_B[f(t)] = \int_{-\infty}^0 e^{-st}f(t)\,dt + \int_0^\infty e^{-st}f(t)\,dt \tag{13-52}$$

We must find a convergence factor $\sigma = \sigma_1$ for the first integral and a $\sigma = \sigma_2$ for the second. The $\mathcal{L}_B$ transform then exists for $\sigma_1 < \sigma < \sigma_2$.

[1] See B. Van der Pol and H. Bremmer, "Operational Calculus Based on the Two-sided Laplace Integral," Cambridge University Press, London, 1950. Also see J. Truxal, "Automatic Feedback Control System Synthesis," pp. 421–424, McGraw-Hill Book Company, Inc., New York, 1955.

From the above observations we conclude that the $\mathcal{L}$ and $\mathcal{F}$ transforms are distinct and that neither is a generalization of the other. Each has its uses, and though the results of transforming appear the same for certain functions, the two have, in general, quite different properties.

There are relations between $\mathcal{L}$ and $\mathcal{F}$ transforms which can be written down. The proofs follow from changes in variables of integration. The relations are:

If $\displaystyle\int_{-\infty}^{\infty} |f(t)|\, dt$ exists and we write $F(s) \triangleq \mathcal{L}[f(t)]$

$$\begin{aligned}
\mathcal{F}[f(t)] &= F(j\omega) + F(-j\omega) && \text{if } f(t) = f(-t) && (f \text{ even}) \\
\mathcal{F}[f(t)] &= F(j\omega) - F(-j\omega) && \text{if } f(t) = -f(-t) && (f \text{ odd})
\end{aligned} \qquad (13\text{-}53)$$

Also

$$\mathcal{L}[f(t)] = \tfrac{1}{2}\{\mathcal{F}[f_1(t)] + \mathcal{F}[f_2(t)]\}_{j\omega=s} = F(s) \qquad (13\text{-}54)$$

where

$$\begin{aligned}
f_1(t) &= f_2(t) = f(t) && t > 0 \\
f_1(-t) &= f_1(t) \\
f_2(-t) &= -f_2(t)
\end{aligned}$$

The formulas (13-53) are useful because $\mathcal{L}$ transforms are more extensively tabulated than $\mathcal{F}$ transforms. If $f(t)$ is neither even $[f(t) = f(-t)]$ nor odd $[f(t) = -f(-t)]$, it can be written as a sum of even and odd functions (see Prob. 13-11).

PROBLEMS

13-1. Verify (3-9) and (3-10).

13-2. Derive expressions (13-43).

13-3. Apply a sinusoid suddenly to a system with transfer function $1/(s + \alpha)(s + \gamma)$, and find the response by the $\mathcal{L}$ method. Compare the result with the steady-state response obtained by the Fourier-transform method.

13-4. Verify the relations in Table 13-1.

13-5. Verify relations (13-53) and (13-54).

13-6. Obtain the result (13-5) using (13-53).

13-7. Show that $\mathcal{L}_B[f(t)] = F(s) + F(-s)$ where $F(s) \triangleq \mathcal{L}[f(t)]$.

13-8. Show from Prob. 13-7 that $\mathcal{L}_B[K]$, K a constant, leads to difficulties. Discuss.

13-9. Verify (13-48) by (13-53).

13-10. Find $\mathcal{L}_B[e^{-a|t|}]$.

13-11. Suppose that $f(t) = f_{\text{even}} + f_{\text{odd}}$. Show that $f_{\text{even}} = [f(t) + f(-t)]/2$, $f_{\text{odd}} = [f(t) - f(-t)]/2$.

13-12. Use (13-53) to find the Fourier transform of:

    a. $e^{-a|t|}$

    b. $t^2 e^{-a|t|}$

    c. $\begin{cases} \sin \omega_1 t & \omega|t| < \pi \\ 0 & \text{elsewhere} \end{cases}$

    d. $\begin{cases} 1 & -1 \le t \le 0 \\ 1 - t & 0 \le t \le 1 \\ 0 & \text{elsewhere} \end{cases}$

    Hint: use results of Prob. 13-11.

13-13. Find the $\mathfrak{F}$ transforms of the derivatives of

    a. $\sin \omega_0 t$

    b. K, a constant

Discuss your results.

13-14. Prove Parseval's relation:

$$\int_{-\infty}^{\infty} f(t)g(t)\ dt = \int_{-\infty}^{\infty} F(\omega)G(-\omega)\ df$$

CHAPTER 14

SYSTEMS WITH RANDOM INPUTS:
THE DIRECT DESCRIPTION

Sometimes a system is subjected to an input which cannot be described completely *before* the time of application. We call such an input *random,*[1] implying uncertainty about its exact nature. If we observe such a function over a number of experiments, we may be able to find out some general facts about it—its average value, for example—and the results of these experiments help us predict what to expect in the future.

Because of the uncertainty in predicting the nature of a random input, we make use of probability and statistics in studying its effect on a system. Many physical problems involve the use of these techniques. Turbulence in aerodynamics and noise in communications and radar systems are examples. In this chapter and the one following, we shall discuss some of the methods for describing random inputs and their effects on linear systems.

We shall discuss two types of description of a random process:

1. Methods based on the direct examination of the input as a function of time.

2. Methods based on examination of the frequency components of the input.

These two ways are analogous to the description of a nonrandom function by transient and frequency methods. We shall discuss the methods of direct examination in this chapter. The frequency methods and applications to systems will be the subject of Chap. 15.

14-1. Probability

We assign a number between 0 and 1 to the chance that an event will occur. If the event is certain, the number is 1, and if impossible, the number is 0. We call this number the *probability* of occurrence.

An example of an event might be the voltage at the output of a noisy receiver exceeding a certain value at some time of observation. If we observe this voltage for a period of T sec and note that the voltage was

---

[1] Sometimes the word *stochastic* is used to describe a random process which is a function of time.

above the prescribed value for t_1 sec, we say that the probability of the event is approximately t_1/T.

Let us talk about two events, A and B. In a hypothetical experiment, suppose we have observed that

A only	occurred n_1 times
B only	occurred n_2 times
A and B	occurred n_3 times
Neither A nor B	occurred n_4 times

We suppose that the total number of trials n was a very large number and write

$$n_1 + n_2 + n_3 + n_4 = n \tag{14-1}$$

If we write the probability of A occurring as $Pr(A)$,[1] then

$$Pr(A) = \frac{n_1 + n_3}{n} \tag{14-2}$$

That is, $Pr(A)$ is the number of times A occurred divided by the number of trials. Similarly,

$$Pr(B) = \frac{n_2 + n_3}{n} \tag{14-3}$$

The event that either A or B or both occur is written $A + B$, and the probability of this event is given by

$$Pr(A + B) = \frac{n_1 + n_2 + n_3}{n} \tag{14-4}$$

The event that both A and B occur is written AB, and its probability is given by

$$Pr(AB) = \frac{n_3}{n} \tag{14-5}$$

The probability that A occurs, given that B has occurred, is called a *conditional probability* and is written $Pr(A/B)$. We have

$$Pr(A/B) = \frac{n_3}{n_2 + n_3} \tag{14-6}$$

We can now establish some of the laws of probability.

From (14-2) through (14-5) we can write

$$Pr(A + B) = Pr(A) + Pr(B) - Pr(AB) \tag{14-7}$$

If A and B are mutually exclusive, then they cannot occur together and

---

[1] We shall use the notation $Pr(\ )$ to mean the probability of the event in the parentheses.

$Pr(AB) = 0$.   In this case the probability of A or B is simply the sum of the individual probabilities.

$$Pr(A + B) = Pr(A) + Pr(B) \qquad A \text{ and } B \text{ mutually exclusive} \qquad (14\text{-}8)$$

From (14-6), (14-3), and (14-5), we write

$$Pr(A/B)\, Pr(B) = \frac{n_3}{n_2 + n_3} \cdot \frac{n_2 + n_3}{n} = Pr(AB) \qquad (14\text{-}9)$$

If A and B are independent, then $Pr(A/B) = Pr(A)$, since A does not depend on B.   In this case, from (14-9),

$$Pr(A)\, Pr(B) = Pr(AB) \qquad A,\, B \text{ independent} \qquad (14\text{-}10)$$

The probability that an event will *not* occur is $1 - Pr$ [event occurs]. This follows from the fact that the sum of probabilities of the two mutually exclusive events of occurrence and nonoccurrence must be unity, since one or the other is certain.   In our example from (14-4) and (14-1)

$$1 - Pr(A + B) = 1 - \frac{n_1 + n_2 + n_3}{n} = \frac{n_4}{n}$$

$$= Pr \text{ [neither } A \text{ nor } B \text{ occurs]}$$
$$(14\text{-}11)$$

Example.   Suppose an experiment consists of tossing two coins simultaneously.   Let us find the probabilities associated with the events: H at least one head occurs, T at least one tail occurs.

The four possible outcomes of the experiment are

$$\begin{array}{ll} \text{head} & \text{head} \\ \text{tail} & \text{tail} \\ \text{head} & \text{tail} \\ \text{tail} & \text{head} \end{array}$$

from which we conclude that $Pr(H) = Pr(T) = \frac{3}{4}$.

Also, we observe that $Pr(HT) = \frac{1}{2}$ and $Pr(H/T) = \frac{2}{3}$.   Now, using results previously derived,

$$\begin{aligned} Pr(H + T) &= \tfrac{3}{4} + \tfrac{3}{4} - \tfrac{1}{2} = 1 \\ Pr(H/T)\, Pr(T) &= \tfrac{2}{3} \times \tfrac{3}{4} = \tfrac{1}{2} = Pr(HT) \end{aligned} \qquad (14\text{-}12)$$

In this example, events H and T are neither mutually exclusive (since there are two coins) nor independent.

14-2. Probability Density

In many physical problems, an experiment involves the observation of a continuous function where the events are no longer discrete.   For example, a random function x might have a history[1] like that shown in Fig. 14-1.

Since there are infinitely many values that the random function can have (like $x = x_0$), the probability of finding x at a particular value at a given time is zero. We can, however, talk about the probability of finding x within a band between x and $x + dx$. We *define* this probability in terms of the *probability density* $p(x)$:

$$p(x_0)\ dx \triangleq Pr[x_0 \leq x \leq x_0 + dx] \qquad dx \to 0 \qquad (14\text{-}13)$$

Since we can have only one value of x per observation, the events $x = x_1$, $x = x_2$, etc., are mutually exclusive. The probability of finding x equal

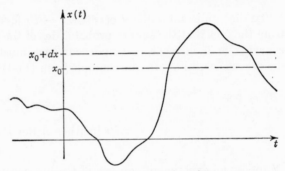

Fig. 14-1. Random function.

to any value between a and b is by (14-8) the sum of the probabilities of finding it in any one of strips that make up the interval. In the limit, as the strips become narrow, the sum becomes the integral:

$$Pr[a \leq x \leq b] = \int_a^b p(x)\ dx \qquad (14\text{-}14)$$

Since we are certain to find x *somewhere*, we can write

$$\int_{-\infty}^{\infty} p(x)\ dx = 1 \qquad (14\text{-}15)$$

When we are considering more than one event for a continuous process, we extend the probability density definition:

$$Pr[x_1 \leq x \leq x_1 + dx \text{ and } y_1 \leq y \leq y_1 + dy] \triangleq p(x_1,y_1)\ dx\ dy$$
$$dx \to 0 \qquad dy \to 0 \qquad (14\text{-}16)$$

---

[1] It is convenient for our purposes to think of the random function x as a function of time. We can, however, generate continuous random functions which do not depend on time. For example, a game in which a coin is tossed at a line drawn on the ground can be used to generate distances from coin to line which can take on any value within reasonable limits.

Here x and y may denote values observed for different processes, or events observed at different times for the same process. If x and y are independent events with probability densities p_x and p_y, then by (14-10)

$$p(x_1, y_1) = p_x(x_1) p_y(y_1) \qquad x, y \text{ independent} \qquad (14\text{-}17)$$

Equation (14-17) is a defining property of independent events.

From (14-16), the probability of finding x in a strip regardless of the value of y is

$$Pr[a \le x \le b \text{ and } -\infty \le y \le \infty] = \int_a^b dx \int_{-\infty}^{\infty} dy\, p(x,y) \qquad (14\text{-}18)$$

from which it follows that

$$p_1(x) = \int_{-\infty}^{\infty} p(x,y)\, dy \qquad (14\text{-}19)$$

where $p_1(x)$ is the probability density for x alone. The density for a single variable from a collection of variables is called a *marginal probability density*.

Example 1. A probability density function which occurs very frequently in physical problems is the *Gaussian* or *normal* function[1] defined by

$$p(x) = \frac{1}{\sigma \sqrt{2\pi}} \exp\left[-\frac{x^2}{2\sigma^2}\right] \qquad (14\text{-}20)$$

and illustrated in Fig. 14-2.

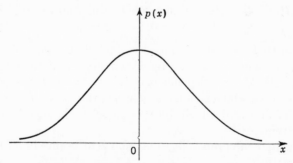

FIG. 14-2. Gaussian probability density.

An interpretation of $p(x)$ in Fig. 14-2 is that the random variable x is more likely to be found in a strip near the origin than anywhere else.

A Gaussian density function when two variables x and y are involved is

$$p(x,y) = \frac{1}{2\pi\sigma_1\sigma_2 \sqrt{1-\rho^2}}\, e^{-\frac{1}{2(1-\rho^2)}\left(\frac{x^2}{\sigma_1^2} - \frac{2\rho xy}{\sigma_1\sigma_2} + \frac{y^2}{\sigma_2^2}\right)} \qquad (14\text{-}21)$$

When x and y are independent $\rho = 0$.

[1] Also discussed in Sec. 2-4.

Example 2.   Let us determine the probability density for the random-phase square wave in Fig. 14-3.   In this case, the function has only two equally likely values, a and $-a$, so that $p(x)$ is concentrated at these points in the form of impulse functions, each with area of $\frac{1}{2}$.

$$p(x) = \frac{1}{2}\delta(x + a) + \frac{1}{2}\delta(x - a) \qquad (14\text{-}22)$$

The density function is illustrated in Fig. 14-4.

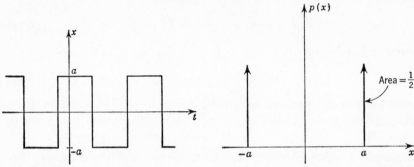

FIG. 14-3. Square wave.

FIG. 14-4. Probability density for square wave.

Example 3 : Poisson Process.   Let us consider a situation in which the probability of occurrence of an event in a short time interval is proportional to the length of the interval:

$$P(\Delta t) \triangleq Pr \text{ [event occurs between } t \text{ and } t + \Delta t]$$
$$= \beta \, \Delta t \qquad \Delta t \to 0 \qquad (14\text{-}23)$$

The condition that $\Delta t \to 0$ is imposed to exclude the possibility of more than one event occurring in Δt.   We also assume that the events of one interval of length Δt are independent of those of all other intervals.

We call a process which generates events in this way a *Poisson* process. An example of an event which fits this model is the disintegration of a radioactive atom.[1]

Let us now find the probability that no event occurs in an interval of T sec.   We write

$$P_0(T) \triangleq Pr \text{ [no event occurs between } t = 0 \text{ and } t = T] \quad (14\text{-}24)$$

Now since the events are assumed independent, we can use (14-10) and write

$$Pr \text{ [no event between } 0 \text{ and } T + \Delta T] = Pr \text{ [no event between } 0 \text{ and } T]$$

$$\times Pr \text{ [no event between } T \text{ and } T + \Delta T] \quad (14\text{-}25)$$

---

[1] For further discussion see, for example, H. Cramér, "Mathematical Methods of Statistics," p. 205, Princeton University Press, Princeton, N.J., 1951.

which becomes, using (14-23), (14-24), and (14-11),

$$P_0(T + \Delta T) = P_0(T)[1 - P(\Delta T)]$$
$$= P_0(T)(1 - \beta \, \Delta T) \tag{14-26}$$

or $$\frac{P_0(T + \Delta T) - P_0(T)}{\Delta T} = -\beta P_0(T) \tag{14-27}$$

When we pass to the limit $\Delta T \to 0$, this becomes

$$\frac{dP_0(T)}{dT} + \beta P_0(T) = 0 \tag{14-28}$$

The initial condition is

$$P_0(0) = Pr \text{ [no event occurs in zero time]} = 1 \tag{14-29}$$

Transforming (14-28) with $\mathcal{L}[P_0(T)] \triangleq \hat{P}_0(s)$,

$$s\hat{P}_0(s) + \beta \hat{P}_0(s) = 1$$
$$\hat{P}_0(s) = \frac{1}{s + \beta}$$
$$P_0(T) = e^{-\beta T} \tag{14-30}$$

Example 4.   Next, let us find the probability density function for the intervals generated by the Poisson process.   If we define

$$p_1(T_0) \, dT \triangleq Pr \text{ [elapsed time between two successive events lies}$$
$$\text{between } T_0 \text{ and } T_0 + dT] \tag{14-31}$$

then from (14-23) and (14-30)

$$p_1(T_0) \, dT = Pr \text{ [no event occurs in } T_0 \text{ sec]}$$
$$\times Pr \text{ [event occurs between } T_0 \text{ and } T_0 + dT]$$
$$= P_0(T_0)\beta \, dT \tag{14-32}$$

and the density function for the intervals is

$$p_1(T) = \beta e^{-\beta T} \tag{14-33}$$

14-3. Averages

The idea of a time average is a familiar one.   For a function $x(t)$, whether it is random or not, the time average $\bar{x}$ is given by[1]

$$\bar{x} = \lim_{T \to \infty} \frac{1}{T} \int_0^T x(t) \, dt \tag{14-34}$$

[1] The average can also be taken in both directions from the origin:

$$\bar{x} = \lim_{T \to \infty} \frac{1}{2T} \int_{-T}^T x(t) \, dt$$

The time average of a function of x is

$$\overline{f(x)} = \lim_{T \to \infty} \frac{1}{T} \int_0^T f[x(t)] \, dt \qquad (14\text{-}35)$$

We often average the function $f(x) = x^2$, for example, since if x is a voltage, $\overline{x^2}$ is the average power dissipated by a 1-ohm resistor when x is impressed across it.

Another kind of average is called the *ensemble* average. Suppose we consider an experiment in which we observe the outputs x_a, x_b, . . . of a number of random-function generators as shown in Fig. 14-5.

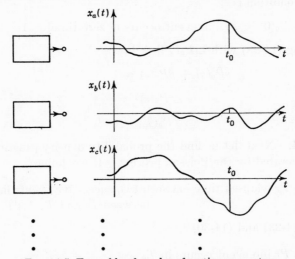

FIG. 14-5. Ensemble of random-function generators.

We can find time averages $\bar{x}_a$, $\bar{x}_b$, etc., but also we can find an average over the ensemble by considering the outputs at time t_0 and forming the quotient

$$\frac{x_a(t_0) + x_b(t_0) + \cdots + x_i(t_0)}{n} \qquad (14\text{-}36)$$

where the outputs of n of the generators have been considered. If we now increase the number of generators n and consider bands of width dx in the output, we can define the ensemble average $\widetilde{x(t_0)}$ as

$$\widetilde{x(t_0)} = \lim_{\substack{n \to \infty \\ dx \to 0}} \frac{n_1 x_1 + n_2 x_2 + \cdots + n_i x_i + \cdots}{n} \qquad (14\text{-}37)$$

where n_i is the number of outputs falling in the band between x_i and $x_i + dx$. We now observe that as n (which is the sum of the n_i) gets large and the dx small, the quotient n_i/n is the probability of finding the value x_i in a strip of width dx:

$$\lim_{n \to \infty} \frac{n_i}{n} = p(x_i)\, dx \qquad (14\text{-}38)$$

Now (14-37) becomes the sum over all the bands of width dx

$$\widetilde{x(t_0)} = \Sigma x_i p(x_i)\, dx \qquad (14\text{-}39)$$

which becomes in the limit as $dx \to 0$ and $n \to \infty$ the integral over all possible values of x,

$$\widetilde{x(t_0)} = \int_{-\infty}^{\infty} x p(x)\, dx \qquad (14\text{-}40)$$

This can be generalized to the ensemble average of a function of x:

$$\widetilde{f[x(t_0)]} = \int_{-\infty}^{\infty} f(x) p(x)\, dx \qquad (14\text{-}41)$$

Stationary and Ergodic Processes. We now define two terms which are useful in describing a random process:

1. A *stationary* process $x(t)$ is one having statistics which do not change with time. If its probability density and all higher-order densities are independent of time, we call the process *strictly stationary*. If the average of the product[1] $x(t)x(t + \tau)$ depends only on τ, we say the process is *wide-sense stationary*.

2. An *ergodic* process is a random process for which time and ensemble averages are equal. (All ergodic processes are stationary, but not necessarily vice versa.)

If the function generators in Fig. 14-5 do not change their characteristics in time, each will generate a stationary process. If, in addition, the process is ergodic, then the result obtained from the ensemble average (14-41) will be the same as the time average (14-35).

The problem of determining whether a physical process is ergodic or not is difficult, since our definitions require an observation of either an infinite ensemble of systems or of one system for an infinite time. The determination must be based on approximate data. The following example illustrates the problem.

A gun is fixed to a rigid frame and fired at a given frequency. The miss distance as a function of time might be as shown in Fig. 14-6. The sources of the misses might be

1. Difference in the rounds—balance, charge, etc.
2. Wind, air density, etc.
3. Erosion of the barrel.

[1] This average product is the *correlation function* which we shall discuss in Sec. 14-5.

If these sources do not change with time, then we could examine either data like Fig. 14-6 or the results of the single firing of a large number of guns and expect the same average.   On the other hand, barrel erosion might be large enough to make the miss-distance plot look like Fig. 14-7.

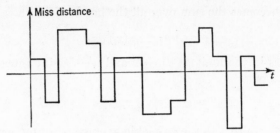

FIG. 14-6. Gun-firing experiment.

Evidently the process is no longer stationary.   In fact, a time average would have little meaning.   Now the ensemble average must be used to give, say, the mean square miss as a function of rounds fired.

In spite of the apparent difficulty in tying down the nature of a general process, it is usually possible in a given situation to say whether a process is ergodic or not.   A great many problems of interest can be considered ergodic, and it is very useful, as we shall see, in related theoretical work to make use of this fact.   Then time and ensemble averages can be equated.

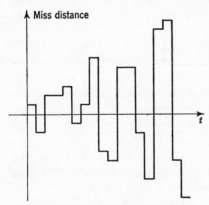

FIG. 14-7. Effect of barrel erosion.

The averages, both time and ensemble, are useful in describing a random process.   Besides physical interpretations, such as d-c value for $\bar{x}$ and average power for $\overline{x^2}$ if x is a voltage, the averages are measures of the shape of probability density functions.   The first moment $\tilde{x}$ is called the *mean* and from (14-40) would be the center of gravity of a thin rod of unit mass with density $p(x)$.   The second moment $\widetilde{x^2}$ would be the moment of inertia about the origin of the same body:

$$\widetilde{x^2} = \int_{-\infty}^{\infty} x^2 p(x) \, dx \qquad (14-42)$$

The second moment about the mean $\overline{(x - \tilde{x})^2}$ is called the *variance* and is a measure of the spread of the density function.   The positive square root of the variance is called *standard deviation*, usually represented by the symbol σ. The σ in the formula (14-20) for the Gaussian function is the standard deviation for that function.

Higher moments are used to describe the density function, but the most widely used are the first and second.[1]

Example. In considering the intervals generated by the Poisson process in Sec. 14-2 we found the probability density function of their lengths $p_1(T)$. Evidently, from (14-38) this can be defined in terms of an experiment. In this experiment we would observe a large number N of intervals and write

$$p_1(T_0)\, dT = \frac{\text{No. of intervals } n_0 \text{ with lengths between } T_0 \text{ and } T_0 + dT}{\text{No. of intervals observed}}$$

$$= \frac{n_0}{N} \qquad N \to \infty \tag{14-43}$$

If we assume that the process is ergodic, then we can observe intervals generated by a single source for a long time *or* we can observe simultaneously the outputs of an ensemble of sources.

Let us now find the probability that a point chosen at random falls in an interval of length between T_0 and $T_0 + dT$ sec. We define

$$p_2(T_0)\, dT \triangleq Pr \,[\text{point falls in interval of length between } T_0 \text{ and}$$
$$T_0 + dT]$$

$$= \frac{\left(\begin{array}{c}\text{time occupied by intervals with}\\ \text{lengths between } T_0 \text{ and } T_0 + dT\end{array}\right)}{\text{total time of observation}} \tag{14-44}$$

Now we write

$$p_2(T_0)\, dT = \frac{n_0 T_0}{(t_1 + t_2 + \cdots + t_N)} = \frac{n_0 T_0}{N\,\dfrac{(t_1 + t_2 + \cdots + t_N)}{N}} \tag{14-45}$$

where t_1, t_2, etc., are the lengths of all the intervals observed. Next, using (14-37) we recognize that the mean interval length is

$$\widehat{\widehat{T}} = \frac{(t_1 + t_2 + \cdots + t_N)}{N} \qquad N \to \infty \tag{14-46}$$

and write (14-45) as

$$p_2(T_0)\, dT = \frac{n_0 T_0}{N\,\widehat{\widehat{T}}} = \frac{T_0}{\widehat{\widehat{T}}}\, p_1(T_0)\, dT \tag{14-47}$$

---

[1] A discussion of other moments and of probability and statistics in general will be found in Cramér, *op. cit.*, and N. Arley and K. R. Buch, "Introduction to the Theory of Probability and Statistics," John Wiley & Sons, Inc., New York, 1950.

for $N \to \infty$. Now from (14-40) and (14-33)

$$\widetilde{T} = \int_{-\infty}^{\infty} Tp_1(T)\, dT = \beta \int_{0}^{\infty} Te^{-\beta T}\, dT = \frac{1}{\beta} \qquad (14\text{-}48)$$

so that for the Poisson process

$$p_2(T) = \beta^2 T e^{-\beta T} \qquad (14\text{-}49)$$

14-4. Characteristic Function

An average of special interest is the mean value of the function $\exp[j\lambda x]$:

$$\phi(\lambda) = \widetilde{e^{j\lambda x}} = \int_{-\infty}^{\infty} e^{j\lambda x} p(x)\, dx \qquad (14\text{-}50)$$

called the *characteristic function*. We see that $\phi(\lambda)$ is a Fourier transform from (13-1):

$$\phi(-\lambda) = \mathfrak{F}[p(x)] \qquad (14\text{-}51)$$

and that

$$p(x) = \mathfrak{F}^{-1}[\phi(-\lambda)] = \frac{1}{2\pi} \int_{-\infty}^{\infty} e^{-j\lambda x} \phi(\lambda)\, d\lambda \qquad (14\text{-}52)$$

To illustrate one use of the characteristic function, let us find the probability density of the sum x of two independent random variables x_1 and x_2,

$$x = x_1 + x_2 \qquad (14\text{-}53)$$

The characteristic function for the sum is

$$\phi(\lambda) = \widetilde{e^{j\lambda x}} = \widetilde{e^{j\lambda(x_1+x_2)}} \qquad (14\text{-}54)$$

Now the average value of a function of two variables can be expressed in terms of the probability density function (14-16):

$$\widetilde{f(x_1,x_2)} = \int_{-\infty}^{\infty} dx_1 \int_{-\infty}^{\infty} dx_2\, f(x_1,x_2) p(x_1,x_2) \qquad (14\text{-}55)$$

The derivation of (14-55) follows the argument leading to (14-41) for a single variable. In the case of the characteristic function, we have then

$$\phi(\lambda) = \widetilde{e^{j\lambda(x_1+x_2)}} = \int_{-\infty}^{\infty} dx_1 \int_{-\infty}^{\infty} dx_2\, e^{j\lambda(x_1+x_2)} p(x_1,x_2) \qquad (14\text{-}56)$$

But x_1 and x_2 were assumed independent, so that from (14-17)

$$p(x_1,x_2) = p_1(x_1) p_2(x_2) \qquad (14\text{-}57)$$

and (14-56) becomes

$$\phi(\lambda) = \int_{-\infty}^{\infty} dx_1\, e^{j\lambda x_1} p_1(x_1) \int_{-\infty}^{\infty} dx_2\, e^{j\lambda x_2} p_2(x_2) \qquad (14\text{-}58)$$

so that we have finally for independent variables

$$\phi(\lambda) = \phi_1(\lambda)\phi_2(\lambda)$$
$$\phi_1(\lambda) = \widetilde{e^{jx_1\lambda}} \qquad (14\text{-}59)$$
$$\phi_2(\lambda) = \widetilde{e^{jx_2\lambda}}$$

Now, returning to (14-52), we have

$$p(x_1 + x_2) = \mathfrak{F}^{-1}[\phi_1(-\lambda)\phi_2(-\lambda)] \qquad (14\text{-}60)$$

But the convolution integral (13-43) is

$$\mathfrak{F}^{-1}[F_1(\omega)F_2(\omega)] = \int_{-\infty}^{\infty} f_1(t_1)f_2(t - t_1)\, dt_1 \qquad (14\text{-}61)$$

Now from (14-52), $\mathfrak{F}^{-1}[\phi_1(-\lambda)] = p_1(x)$, etc., so (14-60) becomes

$$p(x) = p(x_1 + x_2) = \int_{-\infty}^{\infty} p_1(t_1)p_2(x - t_1)\, dt_1 \qquad (14\text{-}62)$$

where p_1 and p_2 are the probability densities for x_1 and x_2.

14-5. Correlation Function

The average value of the product of a function of time with the same function displaced τ sec is called the *correlation function:*[1]

$$R(\tau) \triangleq \overline{x(t)x(t + \tau)} = \lim_{T \to \infty} \frac{1}{T} \int_0^T x(t)x(t + \tau)\, dt \qquad (14\text{-}63)$$

This function is particularly important because it forms the link with the frequency-component methods of description.

For an ergodic process, we can also write $R(\tau)$ as an ensemble average:[2]

$$R(\tau) = \widetilde{x(t)x(t + \tau)} \qquad (14\text{-}64)$$

Some of the properties of $R(\tau)$ for an ergodic process $x(t)$ are:
1. $R(0)$ *is the mean square value of* $x(t)$: $R(0) = \overline{x^2}$.
2. $R(\tau)$ *is an even function:* $R(\tau) = R(-\tau)$.
3. *The magnitude of* $R(\tau)$ *when* $\tau = 0$ *cannot be exceeded for any other value of* τ: $R(0) \geq |R(\tau)|$.

---

[1] This is sometimes called the autocorrelation function to distinguish it from the average of the product of two different functions $\overline{f_1(t)f_2(t + \tau)}$ called the cross-correlation function.

[2] When $R(\tau)$ is written as an ensemble average, it is sometimes called the *covariance function.*

These properties can be proved as follows:

1. From (14-63)

$$R(0) = \lim_{T \to \infty} \frac{1}{T} \int_0^T x^2(t) \, dt = \overline{x^2} \tag{14-65}$$

2. Because $x(t)$ is assumed stationary, the expression (14-63) is independent of the choice of origin and

$$R(\tau) = \overline{x(t)x(t + \tau)} = \overline{x(t - \tau)x(t)} = R(-\tau) \tag{14-66}$$

3. Let us write the average of the squared quantity which is always positive:

$$\overline{[x(t) \pm x(t + \tau)]^2} \geq 0 \tag{14-67}$$

Expanding, we have

$$\overline{x(t)^2} + \overline{x(t + \tau)^2} \pm \overline{2x(t)x(t + \tau)} \geq 0$$
$$R(0) + R(0) \pm 2R(\tau) \geq 0 \tag{14-68}$$

from which we get the result

$$R(0) \geq \mp R(\tau) \tag{14-69}$$

or
$$R(0) \geq |R(\tau)| \tag{14-70}$$

Example 1. Let us find the correlation function of the random-phase sine wave

$$x(t) = \sin (\omega_1 t + \phi) \tag{14-71}$$

We write from (14-63)

$$R(\tau) = \overline{x(t)x(t + \tau)} = \lim_{T \to \infty} \frac{1}{T} \int_0^T \sin (\omega_1 t + \phi) \sin [\omega_1(t + \tau) + \phi] \, dt \tag{14-72}$$

Now $x(t)$ is periodic, so the integral need be taken over only one period:

$$R(\tau) = \frac{\omega_1}{2\pi} \int_0^{2\pi/\omega_1} \sin (\omega_1 t + \phi) \sin (\omega_1 t + \omega_1 \tau + \phi) \, dt$$
$$= \frac{\omega_1}{2\pi} \int_0^{2\pi/\omega_1} \sin^2 (\omega_1 t + \phi) \cos \omega_1 \tau \, dt$$
$$= \tfrac{1}{2} \cos \omega_1 \tau \tag{14-73}$$

In finding the correlation function in (14-73) with a time average, no information about the probability density of ϕ was required. Suppose we assume that values of ϕ are equally distributed between $-\pi$ and π. Then

$$p(\phi) = \begin{cases} 1/2\pi & -\pi \leq \phi \leq \pi \\ 0 & \text{elsewhere} \end{cases} \tag{14-74}$$

Now we can write $R(\tau)$ as the ensemble average with respect to ϕ:

$$R(\tau) = \overline{\sin (\omega_1 t + \phi) \sin [\omega_1(t + \tau) + \phi]}$$

$$= \int_{-\infty}^{\infty} p(\phi) \sin (\omega_1 t + \phi) \sin [\omega_1(t + \tau) + \phi] \, d\phi$$

$$= \frac{1}{2\pi} \int_{-\pi}^{\pi} \sin (\omega_1 t + \phi) \sin [\omega_1(t + \tau) + \phi] \, d\phi$$

$$= \frac{1}{2\pi} \int_{-\pi}^{\pi} \sin^2 (\omega_1 t + \phi) \cos \omega_1 \tau \, d\phi = \frac{1}{2} \cos \omega_1 \tau \qquad (14\text{-}75)$$

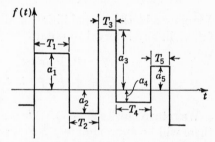

FIG. 14-8. Random square wave.

Since this result agrees with (14-73), we conclude that the $p(\phi)$ chosen makes the process ergodic.

Example 2. Suppose that a random square wave as shown in Fig. 14-8 is somehow generated and that the process is ergodic. The Ts are assumed to be generated by a Poisson process and have probability density given by (14-33)

$$p_1(T) = \beta e^{-\beta T} \qquad T > 0 \qquad (14\text{-}76)$$

The as are assumed to be independent of each other and of the Ts and to have zero mean value.

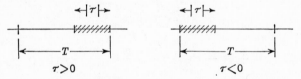

FIG. 14-9. Interval. If t falls in crosshatched region, $t + \tau$ falls outside.

Now for some value of t falling in the ith interval the product $f(t)f(t + \tau)$ will be given by

$$f(t)f(t + \tau) = \begin{cases} a_i^2 & \text{if } t + \tau \text{ falls in } i\text{th interval} \\ a_i a_k & \text{otherwise} \end{cases} \qquad (14\text{-}77)$$

We shall need the probability that t and $t + \tau$ lie in the same interval. Consider an interval as shown in Fig. 14-9. If t falls in the crosshatched region, then $t + \tau$ will fall outside the interval. We write the conditional probability

$P_A = Pr\,[t$ and $t + \tau$ lie in same interval$/t$ falls in interval of length between T and $T + dT]$

$$= \begin{cases} \dfrac{T - |\tau|}{T} & T > |\tau| \\ 0 & T < |\tau| \end{cases} \qquad (14\text{-}78)$$

From (14-9) we write

$P_B = Pr$ [t and $t + \tau$ lie in same interval of length T]

$\quad = P_A \times Pr$ [t falls in interval of length between T and $T + dT$]

$$(14\text{-}79)$$

which becomes, using (14-49) and (14-78),

$$P_B = \begin{cases} \dfrac{T - |\tau|}{T} (\beta^2 T e^{-\beta T}) \, dT & T > |\tau| \\ 0 & T < |\tau| \end{cases} \tag{14-80}$$

We now add the probabilities of all the mutually exclusive events of t falling in any interval to obtain

$$P(\tau) \triangleq Pr \text{ [t and $t + \tau$ lie in same interval]}$$

$$= \beta^2 \int_{|\tau|}^{\infty} (T - |\tau|)e^{-\beta T} \, dT = e^{-\beta|\tau|} \tag{14-81}$$

We return now to (14-77). Over an ensemble of N products $f(t)f(t + \tau)$ there will be approximately $NP(\tau)$ equal to $a_i{}^2$ and $N[1 - P(\tau)]$ equal to $a_i a_k$. Following (14-37), as $N \to \infty$

$$R(\tau) = \widetilde{f(t)f(t + \tau)} = \frac{(n_1 a_1{}^2 + n_2 a_2{}^2 + \cdots)}{N}$$

$$\qquad + \frac{(n_{12} a_1 a_2 + n_{13} a_1 a_3 + \cdots)}{N}$$

$$= P(\tau) \cdot \frac{(n_1 a_1{}^2 + n_2 a_2{}^2 + \cdots)}{NP(\tau)}$$

$$\qquad + [1 - P(\tau)] \cdot \frac{(n_{12} a_1 a_2 + n_{13} a_1 a_3 + \cdots)}{N[1 - P(\tau)]}$$

$$= P(\tau)\widetilde{a_i{}^2} + [1 - P(\tau)]\widetilde{a_i a_k} \tag{14-82}$$

since $(n_1 + n_2 + \cdots) = NP(\tau)$, etc. Now $\widetilde{a_i{}^2}$ is just the mean square value of $f(t)$

$$\widetilde{a_i{}^2} = \widetilde{f(t)^2} \triangleq \widetilde{a^2} \tag{14-83}$$

and $\qquad \widetilde{a_i a_j} = \int_{-\infty}^{\infty} da_i \int_{-\infty}^{\infty} da_j \, a_i a_j p(a_i, a_j) \tag{14-84}$

But the as were assumed independent, so from (14-17)

$$p(a_i, a_j) = p_i(a_i)p_j(a_j) \tag{14-85}$$

and (14-84) becomes

$$\widetilde{a_i a_j} = \int_{-\infty}^{\infty} da_i \, a_i p_i(a_i) \int_{-\infty}^{\infty} da_j \, a_j p(a_j)$$

$$= (\widetilde{a_i})(\widetilde{a_j}) = 0 \tag{14-86}$$

since the as were assumed to have zero mean value. The correlation function is, now, from (14-82) and (14-81)

$$R(\tau) = \widetilde{a^2}e^{-\beta|\tau|} \tag{14-87}$$

The relationship between $R(\tau)$ and the frequency-component description of a random function (the *spectral density*) will be brought out in the next chapter. That a relation of some sort exists can be seen from the following argument. If the random variable has no components of high frequency, it will change slowly with time. A change in $x(t + \tau)$ requiring a large τ would be needed before change in $R(\tau)$ would be noted. This leads us to expect that random functions having narrow (in frequency) spectra around $\omega = 0$ would have broad (in τ) correlation functions.

PROBLEMS

14-1. Discuss an experiment in which two coins are tossed successively. See the example in Sec. 14-1.

14-2. Referring to (14-20), show that $\bar{x} = 0$ and $\overline{x^2} = \sigma^2$.

14-3. Find $\overline{(x - \bar{x})^2}$ in terms of $\bar{x}$ and $\overline{x^2}$.

14-4. A Gaussian random variable x is passed through a perfect linear detector with output y defined by

$$y = \begin{cases} x & x > 0 \\ 0 & x < 0 \end{cases}$$

Sketch the probability density of y.

14-5. Give a proof for (14-41).

14-6. Find $p(x_1 + x_2)$ if x_1 and x_2 are independent random variables with rectangular distributions:

$$p_1(x_1) = p_2(x_2) = \begin{cases} \dfrac{1}{a} & -\dfrac{a}{2} \le x \le \dfrac{a}{2} \\ 0 & \text{elsewhere} \end{cases}$$

14-7. Find the probability density for the difference of two independent random variables.

14-8. Use (14-52) to find the probability density function for $x = \sin(\omega t + \phi)$.

14-9. Give an argument leading to (14-55).

14-10. From (14-50) show that

$$\bar{x} = -j\phi'(0)$$
$$\overline{x^2} = -\phi''(0)$$

14-11. Show that if x_i and x_0 are the stationary input and output of a linear system with impulse response $h(t)$, then

$$\widetilde{x_i(t - \tau_1)x_0(t)} = \int_0^\infty h(\tau)R(\tau - \tau_1)\,d\tau$$

14-12. Show that the d-c value of $x(t)$ is the square root of the average of $R(\tau)$ with respect to τ.

14-13. According to (14-64) the correlation function for an ergodic process can be written

$$R(\tau) = \widetilde{x(t)x(t + \tau)}$$

or, using (14-55),

$$R(\tau) = \int_{-\infty}^{\infty} dx_1 \int_{-\infty}^{\infty} dx_2 \, x_1 x_2 p(x_1, x_2)$$

where

$$x_1 = x(t)$$
$$x_2 = x(t + \tau)$$

Show, using (14-9), that

$$p(x_1, x_2) = p(x_2) \delta[x_1(t) - x_2(t - \tau)]$$

and hence that the double integral reduces to a single integral. Apply this to the verification of (14-75). [Note that if x and y are functionally related, then[1] $p(x)\,dx = p_1(y)\,dy$].

[1] Cramér, *op. cit.*, pp. 292–294.

CHAPTER 15

SYSTEMS WITH RANDOM INPUTS:
THE FREQUENCY DESCRIPTION

It was suggested in the preceding chapter that the correlation function was the link between the direct and frequency descriptions of a random function. The Fourier transform of the correlation function is called the *spectral density*. In this chapter we shall discuss spectral density as a way of describing a random signal and look into the general problem of Fourier integral representation of noise. At the end of the chapter some problems associated with noise applied to systems are discussed.

15-1. Spectral Density

In Chap. 13 we discussed a condition for the existence of a Fourier integral. A sufficient condition for the existence of the transform of $R(\tau)$ is

$$\int_{-\infty}^{\infty} |R(\tau)| \, d\tau < \infty \tag{15-1}$$

If condition (15-1) holds, the Fourier transform of $R(\tau)$ will be bounded. We define the Fourier transform of the correlation function of an ergodic process to be

$$\Phi(\omega) \triangleq \int_{-\infty}^{\infty} R(\tau)e^{-j\omega\tau} \, d\tau \tag{15-2}$$

where $\Phi(\omega)$ is called the *spectral density*. If (15-1) does not hold [e.g., for $R(\tau)$ a sinusoid], the formal application of (15-2) can in some cases still be carried out with impulse functions resulting. We shall say more about this in the next section.

The inversion of (15-2) is (from Chap. 13)

$$R(\tau) = \int_{-\infty}^{\infty} \Phi(\omega)e^{j\omega\tau} \, df \tag{15-3}$$

From (15-3) we note that the mean square value of the random variable is related to the spectral density by

$$\overline{x^2} = R(0) = \int_{-\infty}^{\infty} \Phi(\omega) \, df \tag{15-4}$$

A number of properties of $\Phi(\omega)$ can be deduced from its definition:

1. $\Phi(\omega)$ *is an even function of* ω.

2. $\Phi(\omega)$ *is real.*

3. *The average power dissipated in a 1-ohm resistor by those frequency components of a voltage* $x(t)$ *lying in a band between* f *and* $f + df$ *is* $2\Phi(\omega)\, df.$†

4. $\Phi(\omega)$ *is always positive.*

These properties can be proved as follows:

1. $\Phi(\omega)$ is even. That is, $\Phi(\omega) = \Phi(-\omega)$. This follows from the fact that $R(\tau) = R(-\tau)$ (Sec. 14-5), for if τ is replaced by $-\tau$ in (15-2), we have

$$\Phi(\omega) = - \int_{\infty}^{-\infty} R(-\tau)e^{j\omega\tau}\, d\tau = \int_{-\infty}^{\infty} R(\tau)e^{j\omega\tau}\, d\tau = \Phi(-\omega) \quad (15\text{-}5)$$

2. To show that $\Phi(\omega)$ is real, we make use of the fact that $R(\tau)$ is even and write (15-2)

$$\Phi(\omega) = \int_{-\infty}^{\infty} R(\tau)\,^{-j\omega\tau}\, d\tau = 2 \int_{0}^{\infty} R(\tau)\, \cos\,\omega\tau\, d\tau \quad (15\text{-}6)$$

Since $\Phi(\omega)$ is the integral of a real function, it is itself real.

3. We wish to show that $2\Phi(\omega)\, df$ can be interpreted as the average power dissipated in a 1-ohm resistor by frequency components of $x(t)$ between f and $f + df$.

First we note from (15-4) that the *total* average power of $x(t)$ can be written

$$\overline{x^2} = \int_{0}^{\infty} 2\Phi(\omega)\, df \quad (15\text{-}7)$$

which in turn can be written as the sum of integrals of the form

$$\int_{f_i-\epsilon}^{f_i+\epsilon} 2\Phi(\omega)\, df \quad \epsilon > 0 \quad (15\text{-}8)$$

If each such term corresponds to the average power of frequency components of $x(t)$ near frequency f_i, then the interpretation will be justified.

Suppose now that $x(t)$ has a component $x_1(t)$ of the form

$$x_1(t) = A_1 \sin \omega_1 t \quad (15\text{-}9)$$

Then from (14-73)

$$R_1(\tau) = \frac{A_1{}^2}{2} \cos \omega_1 \tau \quad (15\text{-}10)$$

is the corresponding term in the correlation function. Now from (15-2) and (13-9)[1]

---

† The units of $\Phi(\omega)$ are power/cycle per second.

[1] $R_1(\tau)$ does not satisfy condition (15-1), so that impulse functions in $\Phi(\omega)$ can be expected.

$$\Phi_1(\omega) = \frac{A_1{}^2}{2} \int_{-\infty}^{\infty} \cos \omega_1\tau \, e^{-i\omega\tau} \, d\tau$$

$$= \frac{A_1{}^2}{4} \int_{-\infty}^{\infty} [e^{-i(\omega-\omega_1)\tau} + e^{-i(\omega+\omega_1)\tau}] \, d\tau$$

$$= \frac{A_1{}^2}{4} [\delta(f - f_1) + \delta(f + f_1)] \tag{15-11}$$

The integral of $2\Phi_1(\omega)$ over a narrow interval containing f_1 is

$$\int_{f_1-\epsilon}^{f_1+\epsilon} 2\Phi(\omega) \, df = \frac{A_1{}^2}{2} \qquad \epsilon > 0 \tag{15-12}$$

which is the average power in $x_1(t)$.

An experimental method for measuring spectral density is illustrated in Fig. 15-1. This system will be analyzed in Sec. 15-4. It is important to notice that the order of the squaring and filtering operations cannot be interchanged, since this would introduce components of double the frequency of those present in $x(t)$.

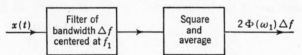

Fig. 15-1. Measurement of spectral density.

4. $\Phi(\omega) \geq 0$, since, from the physical interpretation above, the power dissipated is always positive.

15-2. Fourier Transforms

In Eq. (15-2) we defined the spectral density as the Fourier transform of the correlation function. If condition (15-1) does not hold, this transform will not exist in the strict mathematical sense. The transform may, however, have an integral which exists [the integration of $\Phi(\omega)$ will replace impulse functions by step functions]. The theory of such transforms is called *generalized harmonic analysis*. The following theorem establishes the existence of the integral of spectral density:[1]

Suppose $R(\tau)$ is determined for a stationary process with finite average power. Then the integral

$$\int_{-\infty}^{\infty} R(\tau) \frac{e^{-i\omega\tau} - 1}{-j\tau} \, d\tau = \int_{0}^{\omega} \Phi(\omega_1) \, d\omega_1 \tag{15-13}$$

exists (no impulse functions).[2]

[1] N. Wiener, "The Fourier Integral and Certain of Its Applications," p. 183, Dover Publications, New York, 1933.

[2] For some "pathological" functions (15-13) does not hold, but these can be neglected for most engineering purposes. See N. Wiener, *op. cit.*

Equation (15-13) is simply the integral from 0 to ω of (15-2). This result is an extension of the Fourier-transform theory to correlation functions.

As an example suppose that $x(t) = A \sin(\omega_1 t + \phi)$. Then from (14-73)

$$R(\tau) = \frac{A^2}{2} \cos \omega_1 \tau \qquad (15\text{-}14)$$

which does not satisfy (15-1). However, we have from (15-13)

$$\int_0^\omega \Phi(\omega)\, d\omega = \frac{A^2}{2} \int_{-\infty}^{\infty} \frac{\cos \omega_1 \tau (e^{-j\omega\tau} - 1)}{-j\tau}\, d\tau$$

$$= A_2 \int_0^\infty \frac{\cos \omega_1 \tau \sin \omega \tau}{\tau}\, d\tau$$

$$= \begin{cases} \dfrac{A^2 \pi}{2} & \omega > \omega_1 \\[2mm] 0 & 0 < \omega < \omega_1 \end{cases} \qquad (15\text{-}15)$$

This result is plotted in Fig. 15-2. If we differentiate the step in Fig. 15-2 and make use of the property $\Phi(\omega) = \Phi(-\omega)$, the result is as shown

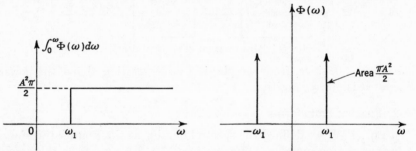

FIG. 15-2. Integrated spectral density of a   FIG. 15-3. Spectral density of sinusoid.
sinusoid.

in Fig. 15-3. The impulse functions indicate that all the power is concentrated at ω_1. The total average power is

$$\overline{x^2} = \int_{-\infty}^{\infty} \Phi(\omega)\, df = \frac{A^2 \pi}{2} \int_{-\infty}^{\infty} [\delta(\omega + \omega_1) + \delta(\omega - \omega_1)]\, \frac{d\omega}{2\pi} = \frac{A^2}{2} \qquad (15\text{-}16)$$

We have shown here that the conditions for existence of a Fourier transform (15-1) can be relaxed for correlation functions. Every correlation function has associated with it a spectral density, possibly containing impulse functions.

15-3. The Fourier Transform of a Random Process

Now let us extend the Fourier-transform theory to another class of functions. A stationary random function $x(t)$ will not satisfy

$$\int_{-\infty}^{\infty} |x(t)|\, dt < \infty \qquad (15\text{-}17)$$

so that its Fourier transform will not be bounded. As in the case of the transform of $R(\tau)$, the transform of $x(t)$ can be taken formally with impulse functions resulting. The transform of $x(t)$ will have an impulse function at every component frequency, so that it is very badly behaved. In fact, except for very simple functions, we cannot show what the transform looks like. However, the integral of the transform is bounded, the integration acting to reduce the impulses to steps.

The following theorem establishes the existence of the Fourier transform of a random process.[1]

Let $x(t)$ be such that $\overline{x(t)^2} < \infty$ and $\widetilde{x(t)x(t+\tau)}$ is independent of time. [x(t) *stationary with finite average power.*] *Then the integral*

$$\mathfrak{F}[x(t)] = \int_{-\infty}^{\infty} x(t)e^{-i\omega t}\,dt \triangleq y(f) \tag{15-18}$$

will not diverge if impulse functions are admitted. [The integral of $y(f)$ *will exist.*]

The inverse of (15-18) can be written[2]

$$x(t) = \int_{-\infty}^{\infty} y(f)e^{i\omega t}\,df \tag{15-19}$$

if impulse functions are used. If $x(t)$ is random, so is $y(f)$.

To see how badly behaved $y(f)$ is, let us average the product $y(f_1)y(f_2)$. Each y corresponds to an $x(t)$ member of an ergodic ensemble of xs. We have from (15-18)

$$\widetilde{y(f_1)y(f_2)} = \overline{\int_{-\infty}^{\infty} e^{-i\omega_1 t_1}x(t_1)\,dt_1 \int_{-\infty}^{\infty} e^{-i\omega_2 t_2}x(t_2)\,dt_2} \tag{15-20}$$

Since $x(t_1)$ and $x(t_2)$ are the only statistical quantities on the right, we can write[3]

$$\widetilde{y(f_1)y(f_2)} = \int_{-\infty}^{\infty} dt_1 \int_{-\infty}^{\infty} dt_2\, \widetilde{x(t_1)x(t_2)}e^{-i\omega_1 t_1}e^{-i\omega_2 t_2} \tag{15-21}$$

Now $x(t)$ is assumed ergodic, so that

$$\widetilde{x(t_1)x(t_2)} = R(\tau) \qquad \tau = t_2 - t_1 \tag{15-22}$$

---

[1] J. L. Doob, "Stochastic Processes," p. 527, John Wiley & Sons, Inc., New York, 1953.

[2] Relation (15-19) is written rigorously as a *Stieltjes integral*

$$x(t) = \int_{-\infty}^{\infty} e^{i\omega t}\,dG(f)$$

where $G(f)$ is the integral of our $y(f)$.

[3] The interchange of averaging and integration is justified as follows:

$$\overline{\int f(t,x)\,dt} = \int_{-\infty}^{\infty} p(x)\,dx \int f(t,x)\,dt = \int dt \int_{-\infty}^{\infty} p(x)f(t,x)\,dx = \int \widetilde{f(t,x)}\,dt$$

and we can write (15-21)

$$\widetilde{y(f_1)y(f_2)} = \int_{-\infty}^{\infty} dt_1 \int_{-\infty}^{\infty} d\tau \, R(\tau)e^{-j\omega_1 t_1}e^{-j\omega_2(t_1+\tau)}$$
$$= \int_{-\infty}^{\infty} d\tau \, R(\tau)e^{-j\omega_2\tau} \int_{-\infty}^{\infty} dt_1 \, e^{-j(\omega_1+\omega_2)t_1} \qquad (15\text{-}23)$$

where we have changed the order of integration. The second integral is the impulse function, so we have finally from (15-2) and (13-9)

$$\widetilde{y(f_1)y(f_2)} = \int_{-\infty}^{\infty} d\tau \, R(\tau)e^{-j\omega_2\tau}\delta(f_1+f_2) = \Phi(\omega_2)\delta(f_1+f_2) \qquad (15\text{-}24)$$

The expression (15-24) shows that the correlation function of y, $\widetilde{y(f_1)y(f_2)}$, has a value different from zero only when $f_1 = -f_2$, and there the value is infinite.

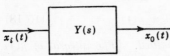

$x_i(t)$ —→ | $Y(s)$ | —→ $x_0(t)$

Fig. 15-4. System with random input.

15-4. Filtered Noise

In this section we shall find the correlation function of the output of a system subjected to a random input. Our system is shown in Fig. 15-4. If we define $y_i(f) \triangleq \mathfrak{F}[x_i(t)]$, we can write the correlation function of the input:

$$R_i(\tau) = \widetilde{x_i(t)x_i(t+\tau)} = \int_{-\infty}^{\infty} y_i(f_1)e^{j\omega_1 t}\,df_1 \int_{-\infty}^{\infty} y_i(f_2)e^{j\omega_2(t+\tau)}\,df_2$$
$$= \int_{-\infty}^{\infty} df_1 \int_{-\infty}^{\infty} df_2 \, \widetilde{y_i(f_1)y_i(f_2)}e^{j(\omega_1+\omega_2)t}e^{j\omega_2\tau} \qquad (15\text{-}25)$$

Now from (15-24) we write

$$R_i(\tau) = \int_{-\infty}^{\infty} df_2 \int_{-\infty}^{\infty} df_1 \, \Phi_i(\omega_2)\delta(f_1+f_2)e^{j(\omega_1+\omega_2)t}e^{j\omega_2\tau}$$
$$= \int_{-\infty}^{\infty} df_2 \, \Phi_i(\omega_2)e^{j\omega_2\tau} \qquad (15\text{-}26)$$

The result (15-26) is the definition of the spectral density of the input $\Phi_i(\omega)$ from (15-3) and serves as a check on the method.

We now do the same thing for the output. The Fourier transform of $x_0(t)$ is (see Sec. 13-2)

$$\mathfrak{F}[x_0(t)] = Y(j\omega)\mathfrak{F}[x_i(t)] = Y(j\omega)y_i(f) \qquad (15\text{-}27)$$

The correlation function of the output is

$$R_0(\tau) = \widetilde{x_0(t)x_0(t+\tau)}$$
$$= \int_{-\infty}^{\infty} Y(j\omega_1)y(f_1)e^{j\omega_1 t}\,df_1 \int_{-\infty}^{\infty} Y(j\omega_2)y(f_2)e^{j\omega_2(t+\tau)}\,df_2$$
$$= \int_{-\infty}^{\infty} df_2 \int_{-\infty}^{\infty} df_1 \, Y(j\omega_1)Y(j\omega_2)\widetilde{y(f_1)y(f_2)}e^{j(\omega_1+\omega_2)t}e^{j\omega_2\tau} \qquad (15\text{-}28)$$

Again using (15-24) we have

$$\overline{x_0(t)x_0(t + \tau)} = \int_{-\infty}^{\infty} df_2 \int_{-\infty}^{\infty} df_1\, Y(j\omega_1)Y(j\omega_2)\Phi_i(\omega_2)\delta(f_1 + f_2)e^{j(\omega_1+\omega_2)t}e^{j\omega_2\tau}$$

$$= \int_{-\infty}^{\infty} df_2\, Y(-j\omega_2)Y(j\omega_2)\Phi_i(\omega_2)e^{j\omega_2\tau} \qquad (15\text{-}29)$$

So finally

$$R_0(\tau) = \int_{-\infty}^{\infty} [|Y(j\omega_2)|^2\Phi_i(\omega_2)]e^{j\omega_2\tau}\, df_2 \qquad (15\text{-}30)$$

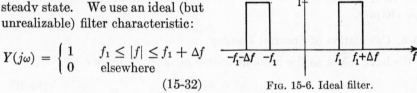

FIG. 15-5. Experimental determination of $\Phi(\omega)$.

Equation (15-30) shows that the spectral density of the output (which is the Fourier transform of the output correlation function) is

$$\Phi_0(\omega) = |Y(j\omega)|^2\Phi_i(\omega) \qquad (15\text{-}31)$$

Example. An example is furnished by the system of Fig. 15-1 which could be used to measure spectral density. The system is shown again in Fig. 15-5, where $x_i(t)$ is an ergodic process, as is $x_0(t)$ in the steady state. We use an ideal (but unrealizable) filter characteristic:

$$Y(j\omega) = \begin{cases} 1 & f_1 \le |f| \le f_1 + \Delta f \\ 0 & \text{elsewhere} \end{cases} \qquad (15\text{-}32)$$

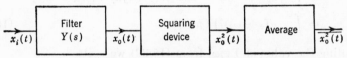

FIG. 15-6. Ideal filter.

$Y(j\omega)$ is plotted in Fig. 15-6. If we write $\mathfrak{F}[x_i(t)] \triangleq y_i(f)$, etc., then by (15-27)

$$y_0(f) = y_i(f)Y(j\omega) \qquad (15\text{-}33)$$

The Fourier transform of the square of $x_0(t)$ is by (13-42)

$$\mathfrak{F}[x_0^2(t)] = \int_{-\infty}^{\infty} y_0(f_2)y_0(f - f_2)\, df_2$$

$$= \int_{-\infty}^{\infty} y_i(f_2)y_i(f - f_2)Y(j\omega_2)Y[j(\omega - \omega_2)]\, df_2 \qquad (15\text{-}34)$$

The output of the system of Fig. 15-5 is $\overline{x_0^2(t)}$, which is the same as $\overline{\overline{x_0^2(t)}}$, since $x_0(t)$ is an ergodic process. Now

$$\mathfrak{F}[\overline{\overline{x_0^2(t)}}] = \overline{\overline{\mathfrak{F}[x_0^2(t)]}} = \int_{-\infty}^{\infty} \overline{\overline{y_i(f_2)y_i(f - f_2)}}\,Y(j\omega_2)Y[j(\omega - \omega_2)]\, df_2$$

$$(15\text{-}35)$$

From (15-24) we write finally

$$\mathfrak{F}[\widetilde{x_0{}^2(t)}] = \int_{-\infty}^{\infty} \Phi_i(\omega_2)\delta(f)Y(j\omega_2)Y[j(\omega - \omega_2)]\,df_2 \tag{15-36}$$

We now take the inverse transform of (15-36)

$$\begin{aligned}
\widetilde{x_0{}^2(t)} &= \int_{-\infty}^{\infty} df\, e^{j\omega t} \int_{-\infty}^{\infty} df_2\, \Phi_i(\omega_2)\delta(f)Y(j\omega_2)Y[j(\omega - \omega_2)] \\
&= \int_{-\infty}^{\infty} df_2 \int_{-\infty}^{\infty} df\, e^{j\omega t}\Phi_i(\omega_2)\delta(f)Y(j\omega_2)Y[j(\omega - \omega_2)] \\
&= \int_{-\infty}^{\infty} df_2\, \Phi_i(\omega_2)|Y(j\omega_2)|^2 \tag{15-37}
\end{aligned}$$

Substituting the value of $Y(j\omega)$ from (15-32) we write

$$\begin{aligned}
\widetilde{x_0(t)^2} &= \int_{-f_1-\Delta f}^{-f_1} \Phi_i(\omega_2)\,df_2 + \int_{f_1}^{f_1+\Delta f} \Phi(\omega_2)\,df_2 \\
&= 2\int_{f_1}^{f_1+\Delta f} \Phi_i(\omega_2)\,df_2 \tag{15-38}
\end{aligned}$$

If we have used a filter with small Δf so that $\Phi(\omega)$ does not change much in the interval, we can take $\Phi(\omega)$ outside the integral and write the approximate relation

$$\widetilde{x_0{}^2} \cong 2\Phi(f_1)\,\Delta f \tag{15-39}$$

This result is the same as the interpretation of spectral density given in Sec. 15-1. It furnishes a good example of the use of the techniques in this chapter.

15-5. Calculation of Spectral Density

We have taken as the basic definition of spectral density

$$\Phi(\omega) \triangleq \mathfrak{F}[R(\tau)] \tag{15-40}$$

In the preceding sections we have developed the physical interpretation of $\Phi(\omega)$ from its properties. The question of the existence of the transform of $R(\tau)$ and of a stationary random variable has been discussed. We have seen that the spectral density of the output of a linear device is simply related to that of the input (15-31). If the spectral density of the output of a device is known, then the total average power at the output is known also by (15-4). This is the basis for design of systems which operate in some optimum fashion in the presence of noise.[1]

The determination of spectral density, since it forms a basis for system design, is an important problem. Several methods will be discussed here.

1. *Spectral density from $R(\tau)$.* If $R(\tau)$ is obtained from experimental data, the Fourier transform can be taken using an approximate analytical

---

[1] See, for example, J. G. Truxal, "Automatic Feedback Control System Synthesis," chap. 8, McGraw-Hill Book Company, Inc., New York, 1955.

expression for $R(\tau)$ or by using an approximating sum for the integral. The expression (15-2) may be written in real form for this purpose:

$$\Phi(\omega) = \int_{-\infty}^{\infty} e^{-j\omega\tau} R(\tau) \, d\tau = 2 \int_0^{\infty} R(\tau) \cos \omega\tau \, d\tau \qquad (15\text{-}41)$$

Example. In Sec. 14-5 we found in (14-87) that the correlation function of a random square wave (Fig. 14-8) was

$$R(\tau) = \widetilde{a^2} e^{-\beta|\tau|} \qquad (15\text{-}42)$$

From this the spectral density can be calculated with (15-41):

$$\begin{aligned}
\Phi(\omega) &= 2\widetilde{a^2} \int_0^{\infty} e^{-\beta\tau} \cos \omega\tau \, d\tau \\
&= 2\widetilde{a^2} \{\mathcal{L}[\cos \omega\tau]\}_{s=\beta} \\
&= \frac{2\widetilde{a^2}\beta}{\omega^2 + \beta^2} \qquad (15\text{-}43)
\end{aligned}$$

where we have used the $\mathcal{L}$-transform table to evaluate the integral. The average power of the square wave is

$$\begin{aligned}
\widetilde{\overline{f(t)^2}} &= \int_{-\infty}^{\infty} \Phi(\omega) \, df = \frac{2\widetilde{a^2}\beta}{2\pi} \int_{-\infty}^{\infty} \frac{d\omega}{\omega^2 + \beta^2} \\
&= \frac{2\widetilde{a^2}\beta}{2\pi} \left(\frac{\pi}{\beta}\right) = \widetilde{a^2} \qquad (15\text{-}44)
\end{aligned}$$

which agrees with $R(0)$ from (15-42).

2. *Direct Measurement of Spectral Density.* The experimental arrangement of Fig. 15-5 can be used.[1]

3. *Direct Calculation of Spectral Density.* A method for determination of $\Phi(\omega)$ directly from $x(t)$ is available. Let us introduce the function

$$A_T(f) \triangleq \int_{-T}^{T} e^{-j\omega t} x(t) \, dt \qquad (15\text{-}45)$$

where we recognize that $A_T(f) \rightarrow y(f)$ as $T \rightarrow \infty$. Let us form

$$\begin{aligned}
\mathcal{F}^{-1}\left[\frac{1}{2T} |A_T(f)|^2\right] &= \frac{1}{2T} \int_{-\infty}^{\infty} e^{j\omega\tau} |A_T(f)|^2 \, df \\
&= \frac{1}{2T} \int_{-\infty}^{\infty} e^{j\omega\tau} \, df \int_{-T}^{T} e^{-j\omega t_1} x(t_1) \, dt_1 \int_{-T}^{T} e^{j\omega t_2} x(t_2) \, dt_2
\end{aligned}$$

$$(15\text{-}46)$$

[1] This is discussed in detail by R. R. Bennett and A. S. Fulton, The Generation and Measurement of Low Frequency Random Noise, *J. Appl. Phys.*, **22**:1187–1191 (1951).

Upon interchange of order of integration we have

$$\mathfrak{F}^{-1}\left[\frac{1}{2T}|A_T(f)|^2\right] = \frac{1}{2T}\int_{-T}^{T} dt_2 \int_{-T}^{T} dt_1\, x(t_1)x(t_2) \int_{-\infty}^{\infty} e^{j\omega(\tau+t_2-t_1)}\,df$$

$$= \frac{1}{2T}\int_{-T}^{T} dt_2 \int_{-T}^{T} dt_1\, x(t_1)x(t_2)\delta(\tau+t_2-t_1)$$

$$= \frac{1}{2T}\int_{-T}^{T} dt_2\, x(t_2+\tau)x(t_2) \qquad (15\text{-}47)$$

We can now take the limit as $T \to \infty$ of both sides of (15-47). The right-hand side $\to R(\tau)$ which is the $\mathfrak{F}^{-1}$ transform of spectral density. We write the result:

$$\lim_{T\to\infty} \mathfrak{F}^{-1}\left[\frac{1}{2T}|A_T(f)|^2\right] = R(\tau) = \mathfrak{F}^{-1}[\Phi(\omega)] \qquad (15\text{-}48)$$

We would be tempted now to write that

$$\lim_{T\to\infty} \frac{1}{2T}|A_T(f)|^2 = \Phi(\omega) \qquad (15\text{-}49)$$

since both have the same inverse Fourier transform. This is *incorrect* even if $\Phi(\omega)$ has a derivative everywhere. A *correct* expression is[1]

$$\lim_{T\to\infty} \int_{f_1}^{f_2} \frac{1}{2T}|A_T(f)|^2\,df = \int_{f_1}^{f_2} \Phi(\omega)\,df \qquad (15\text{-}50)$$

where the limit and integral cannot be interchanged. The reason for this is roughly as follows: $A_T(f)$ approaches $y(f)$ of (15-18) in the limit. As we saw in Sec. 15-3, $y(f)$ is badly behaved. In fact, the integral of $y(f)$ is everywhere discontinuous—being made up of many small step functions—so $y(f)$ consists of a smear of impulse functions. Now whereas the integral of

$$\frac{1}{2T}|A_T(f)|^2$$

approaches

$$\int_{f_1}^{f_2} \Phi(\omega)\,df$$

as the steps become smaller, the functions themselves cannot be equal because of the unbounded behavior of the left side of (15-49). Attempts to evaluate $\Phi(\omega)$ by (15-49) will lead to functions which deviate more and more violently from $\Phi(\omega)$ as T is made large.

The failure of (15-49) is interesting because it is caused by the occurrence of a "pathological" function as a result of a physical problem. Functions which are everywhere discontinuous are generated sometimes by mathematicians as horrible examples to illustrate what can happen if

[1] Doob, *op. cit.*, p. 531.

rigor is not maintained.   The occurrence of such a function in a physical problem emphasizes the possibility of failure of the kind of formal mathematics in this book and should make the user of such techniques cautious. With caution, however, these techniques lead to physical understanding often submerged under rigorous treatment.

An alternate correct form which is useful in analysis for direct calculation of spectral density is

$$\Phi(\omega) = \lim_{T \to \infty} \frac{1}{2T} \widetilde{|A_T(f)|^2} \tag{15-51}$$

To prove this, let us write

$$A_T(\omega) = \int_{-T}^{T} x(t)e^{-j\omega t}\, dt$$
$$= \int_{-T}^{T} dt\, e^{-j\omega t} \int_{-\infty}^{\infty} df_1\, y(f_1)e^{j\omega_1 t} \tag{15-52}$$

where we have replaced $x(t)$ by its Fourier integral representation (15-19). Interchanging order of integration, we have

$$A_T(\omega) = \int_{-\infty}^{\infty} df_1\, y(f_1) \int_{-T}^{T} e^{-j(\omega-\omega_1)t}\, dt$$
$$= \int_{-\infty}^{\infty} df_1\, y(f_1) \frac{e^{j(\omega-\omega_1)T} - e^{-j(\omega-\omega_1)T}}{j(\omega - \omega_1)}$$
$$= \int_{-\infty}^{\infty} df_1\, y(f_1) \frac{2 \sin (\omega - \omega_1)T}{(\omega - \omega_1)} \tag{15-53}$$

Now from (15-52) the conjugate of $A_T(\omega)$ is

$$A_T^*(\omega) = \int_{-T}^{T} x(t)e^{j\omega t}\, dt = A_T(-\omega) \tag{15-54}$$

so

$$\frac{1}{T} |A_T(\omega)|^2 = \frac{1}{T} A_T(\omega) A_T(-\omega)$$
$$= \int_{-\infty}^{\infty} df_1 \int_{-\infty}^{\infty} df_2\, y(f_1)y(f_2) \frac{4 \sin (\omega - \omega_1)T \sin (-\omega - \omega_2)T}{(\omega - \omega_1)(-\omega - \omega_2)T} \tag{15-55}$$

We now take the ensemble average and substitute from (15-24):

$$\frac{1}{2T} \widetilde{|A_T(\omega)|^2} = 2 \int_{-\infty}^{\infty} df_1 \int_{-\infty}^{\infty} df_2\, \widetilde{y(f_1)y(f_2)} \frac{\sin (\omega - \omega_1)T \sin (\omega + \omega_2)T}{(\omega - \omega_1)(\omega + \omega_2)T}$$
$$= 2 \int_{-\infty}^{\infty} df_1 \int_{-\infty}^{\infty} df_2\, \Phi(\omega_1)\delta(f_1 + f_2) \frac{\sin (\omega - \omega_1)T \sin (\omega + \omega_2)T}{(\omega - \omega_1)(\omega + \omega_2)T}$$
$$= 2 \int_{-\infty}^{\infty} df_1\, \Phi(\omega_1) \frac{\sin^2 (\omega - \omega_1)T}{(\omega - \omega_1)^2 T} \tag{15-56}$$

Now the term in the integrand

$$G(\omega) \triangleq 2 \frac{\sin^2 (\omega - \omega_1)T}{(\omega - \omega_1)^2 T} \tag{15-57}$$

is an example of a function from which an impulse function can be generated (see Sec. 3-1).   Figure 15-7 is a sketch of $G(\omega)$.   The integral of $G(\omega)$ with respect to f is unity regardless of the value of T.   As T becomes

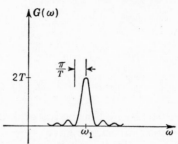

FIG. 15-7. Function which approaches an impulse function.

large, the function becomes a narrow pulse which is very nearly zero everywhere except at $\omega = \omega_1$.   In the limit

$$\lim_{T \to \infty} G(\omega) = \lim_{T \to \infty} 2 \frac{\sin^2 (\omega - \omega_1)T}{(\omega - \omega_1)^2 T} = \delta(f - f_1) \tag{15-58}$$

Now (15-56) becomes

$$\lim_{T \to \infty} \frac{1}{2T} \overline{|A_T(\omega)|^2} = \int_{-\infty}^{\infty} df_1 \, \Phi(\omega_1) \delta(f - f_1) = \Phi(\omega) \tag{15-59}$$

Example.   As an example of direct calculation of spectral density let us return to the random square wave of Fig. 14-8.   We shall use (15-51) to calculate $\Phi(\omega)$.

It will be convenient to use an equivalent expression for A_T.   Since the process is ergodic, the time origin is of no consequence and we can write for large T

$$\frac{1}{2T} |A_T(f)|^2 = \frac{1}{2T} \left| \int_{-T}^{T} e^{-i\omega t} x(t) \, dt \right|^2 \cong \frac{1}{T} \left| \int_{0}^{T} e^{-i\omega t} x(t) \, dt \right|^2$$

$$\triangleq \frac{1}{T} |\hat{A}_T|^2 \tag{15-60}$$

Let us use as the range of integration $N\bar{T}$ where $\bar{T}$ is the average interval length of the square wave.   Now we write

$$\hat{A}_T(f) = \int_0^{N\bar{T}} f(t)e^{-i\omega t}\,dt$$

$$= \left[a_1 \int_0^{T_1} e^{-i\omega t}\,dt + a_2 \int_{T_1}^{T_1+T_2} e^{-i\omega t}\,dt + \cdots \right]$$

$$= \frac{1}{j\omega}\left[a_1(1 - e^{-i\omega T_1}) + a_2(e^{-i\omega T_1} - e^{-i\omega(T_1+T_2)}) + \cdots \right] \quad (15\text{-}61)$$

For large T there will be approximately N such terms.

Let us now form

$$|\hat{A}_T(f)|^2 = \frac{1}{\omega^2}\left[a_1(1 - e^{-i\omega T_1}) + a_2(e^{-i\omega T_1} - e^{-i\omega(T_1+T_2)}) + \cdots \right]$$

$$\times \left[a_1(1 - e^{i\omega T_1}) + a_2(e^{i\omega T_1} - e^{i\omega(T_1+T_2)}) + \cdots \right]$$

$$= \frac{1}{\omega^2}\{[2a_1{}^2(1 - \cos \omega T_1) + 2a_2{}^2(1 - \cos \omega T_2) + \cdots]$$

$$+ \text{[terms with } a_1 a_2 \text{ and all other cross products } a_i a_j$$
$$\text{as factors]}\} \quad (15\text{-}62)$$

Now according to (15-51) we take the ensemble average. Since the as and Ts are independent, the average of the product is the product of averages as in (14-86). Also, since from (14-86) $\widetilde{a_i a_j}$ is zero for $i \neq j$, the terms involving cross products are zero. We now have

$$\widetilde{|\hat{A}_T(f)|^2} = \frac{2}{\omega^2}[\widetilde{a_1{}^2}(1 - \widetilde{\cos \omega T_1}) + \widetilde{a_2{}^2}(1 - \widetilde{\cos \omega T_2}) + \cdots] \quad (15\text{-}63)$$

Now from (14-83)

$$\widetilde{a_1{}^2} = \widetilde{a_2{}^2} = \cdots \triangleq \widetilde{a^2} \quad (15\text{-}64)$$

so (15-63) can be written

$$\widetilde{|\hat{A}_T(f)|^2} = \frac{2\widetilde{a^2}}{\omega^2}[(1 - \widetilde{\cos \omega T_1}) + (1 - \widetilde{\cos \omega T_2}) + \cdots] \quad (15\text{-}65)$$

Similarly

$$(1 - \widetilde{\cos \omega T_1}) = (1 - \widetilde{\cos \omega T_2}) = \cdots = (1 - \widetilde{\cos \omega T_i}) \quad (15\text{-}66)$$

and there are N such terms for large N. We now write

$$\widetilde{|\hat{A}_T(f)|^2} = \frac{2\widetilde{a^2}N}{\omega^2}(1 - \widetilde{\cos \omega T_1}) \quad (15\text{-}67)$$

and $\quad \Phi(\omega) = \lim_{T \to \infty} \frac{1}{T}\widetilde{|\hat{A}_T(f)|^2} = \frac{1}{N\bar{T}}\frac{2\widetilde{a^2}N}{\omega^2}(1 - \widetilde{\cos \omega T_1})$

$$= \frac{2\widetilde{a^2}}{\bar{T}\omega^2}(1 - \widetilde{\cos \omega T_1}) \quad (15\text{-}68)$$

Now using the probability density of lengths from (14-76),

$$\widetilde{\cos \omega T_1} = \int_{-\infty}^{\infty} p_1(T_1) \cos \omega T_1 \, dT_1$$

$$= \beta \int_0^{\infty} e^{-\beta T_1} \cos \omega T_1 \, dT_1$$

$$= \frac{\beta^2}{\omega^2 + \beta^2} \tag{15-69}$$

and from (14-48)

$$\bar{T} = \frac{1}{\beta} \tag{15-70}$$

so finally

$$\Phi(\omega) = \frac{2\widetilde{a^2}}{\omega^2} \beta \left(1 - \frac{\beta^2}{\omega^2 + \beta^2} \right)$$

$$= \frac{2\widetilde{a^2}\beta}{\omega^2 + \beta^2} \tag{15-71}$$

which is the same as the result (15-43) obtained from the correlation function.

15-6. Transient Response to Random Input

When a stationary signal is suddenly applied to a system, a non-stationary output results. In our previous discussion of steady-state conditions we have assumed that sufficient time has elapsed so that the output has become stationary. We shall now discuss the output of a linear system during the transient.

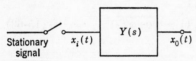

Stationary signal $x_i(t)$ $Y(s)$ $x_0(t)$

FIG. 15-8. System with noise suddenly applied.

We consider the system in Fig. 15-8. The switch is closed at $t = 0$. The output is, using $\mathcal{L}$ transforms where $X_i(s) \triangleq \mathcal{L}[x_i(t)]$, etc.,

$$x_0(t) = \mathcal{L}^{-1}[X_i(s) Y(s)]$$

$$= \int_0^t h(\tau_1) x_i(t - \tau_1) \, d\tau_1 \tag{15-72}$$

where $h(t) \triangleq \mathcal{L}^{-1}[Y(s)]$ is the impulse response of the system. The mean square output is

$$\widetilde{x_0(t)^2} = \int_0^t d\tau_1 \int_0^t d\tau_2 \, \widetilde{x_i(t - \tau_1) x_i(t - \tau_2)} h(\tau_1) h(\tau_2) \tag{15-73}$$

where we have taken the ensemble average of the nonstationary $x_0^2(t)$. The input, being stationary for $t > 0$, has a correlation function independent of time, so we can write

$$\widetilde{x_0(t)^2} = \int_0^t d\tau_1 \int_0^t d\tau_2 \, R(\tau_1 - \tau_2) h(\tau_1) h(\tau_2) \tag{15-74}$$

An example of the use of (15-74) will be given in the next section.

15-7. Examples

Examples illustrating material from the preceding sections are presented here.

Example 1: White Noise. It is often useful to consider noise with constant spectral density. This is a physical impossibility because the average power of such a process would be infinite if $\Phi(\omega) = N_0$,

$$\overline{x^2} = \int_{-\infty}^{\infty} \Phi(\omega)\, df = N_0 \int_{-\infty}^{\infty} df \rightarrow \infty \qquad (15\text{-}75)$$

This is reflected in the correlation function

$$R(\tau) = N_0 \int_{-\infty}^{\infty} e^{j\omega\tau}\, df = N_0\delta(\tau)$$
$$R(0) \rightarrow \infty \qquad\qquad (15\text{-}76)$$

In spite of these apparent difficulties the fact that physical systems do not pass all frequencies means that when white noise is applied to a system, noise with finite average power results.

Another justification for the use of white noise is its occurrence for most engineering purposes in nature. The noise due to thermal agitation of electrons in a conductor, for example, has a constant spectrum out to a frequency of 1.7×10^{13} cps[1] at which frequency $\Phi(\omega)$ has dropped by one-half. The bandwidth, which is predicted by quantum mechanics, is sufficiently above that considered in most applications to make the assumption of white noise reasonable.

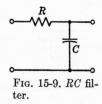

Fig. 15-9. RC filter.

Example 2: White Noise Applied to an RC Filter. If white noise of spectral density N_0 is applied to the system of Fig. 15-8, we have from (15-74) and (15-76)

$$\widetilde{x_0(t)^2} = \int_0^t d\tau_1 \int_0^t d\tau_2\, N_0\delta(\tau_1 - \tau_2)h(\tau_1)h(\tau_2)$$
$$= N_0 \int_0^t d\tau_1\, h^2(\tau_1) \qquad (15\text{-}77)$$

Now for a filter such as shown in Fig. 15-9 the transfer function is

$$Y(s) = \frac{1}{RCs + 1} = \frac{1}{RC}\frac{1}{s + (1/RC)} \qquad (15\text{-}78)$$

and
$$h(t) \triangleq \mathcal{L}^{-1}[Y(s)] = \frac{1}{RC} e^{-(t/RC)} \qquad (15\text{-}79)$$

---

[1] W. R. Bennett, Sources and Properties of Electrical Noise, *Elec. Eng.*, **73**:1001-1008 (November, 1954).

TABLE 15-1. RELATIONS AMONG DESCRIPTIONS

| | Descriptions—Assume | | | |
	$x(t)$	$p(x)$	$\bar{x}$	$\overline{x^2}$
$x(t)$	—	No	No	No
$p(x)$	From Record—See Note 1	—	No	No
$\bar{x}$	$\lim\limits_{T\to\infty} \dfrac{1}{T}\displaystyle\int_0^T x(t)\,dt$	$\displaystyle\int_{-\infty}^{\infty} x p(x)\,dx$	—	No
$\overline{x^2}$	$\lim\limits_{T\to\infty} \dfrac{1}{T}\displaystyle\int_0^T x^2(t)\,dt$	$\displaystyle\int_{-\infty}^{\infty} x^2 p(x)\,dx$	No	—
$\phi(\lambda)$	From $p(x)$	$\displaystyle\int_{-\infty}^{\infty} e^{i\lambda x} p(x)\,dx$	No	No
$p(x,y)$ $x = x(t)$ $y = x(t+\tau)$	From Record—See Note 1	No	No	No
$R(\tau)$	$\lim\limits_{T\to\infty} \dfrac{1}{T}\displaystyle\int_0^T x(t)x(t+\tau)\,dt$	No	No	No
$\Phi(\omega)$	From $R(\tau)$ or $\lim\limits_{T\to\infty} \dfrac{1}{2T}\left| \displaystyle\int_{-T}^{T} e^{-i\omega t} x(t)\,dt \right|^2$	No	No	No

Note 1: From a record of T sec of $x(t)$ we form, as $T \to \infty$ and $\Delta x, \Delta y \to 0$, $p(x_1)\,\Delta x = [\text{time } x(t) \text{ is between } x_1 \text{ and } x_1 + \Delta x]/T$; $p(x_1,y_1)\,\Delta x\,\Delta y = (\text{time } x \text{ is between } x_1 \text{ and } x_1 + \Delta x \text{ and } y \text{ between } y_1 \text{ and } y_1 + \Delta y)/T$.

We now have from (15-77), with $\Phi(\omega) = N_0$ volt^2/cps,

$$\widetilde{x_0^2(t)} = \frac{N_0}{R^2 C^2} \int_0^t e^{-(2\tau_1/RC)}\,d\tau_1 = \frac{N_0}{2RC}\left[1 - e^{-(2t/RC)}\right] \qquad (15\text{-}80)$$

The mean square starts at zero and becomes $N_0/2RC$ in the steady state.

Example 3: Filtered Gaussian White Noise. If the input of a linear system has a Gaussian probability density, so does the output.[1] If we assume for convenience that the input contains no direct current $(\overline{x_i} = 0)$, then the mean square of the output is σ^2 and probability density is, by (14-20),

$$p(x_0) = \frac{1}{(2\pi)^{1/2}\sigma} e^{-(x_0^2/2\sigma^2)} \qquad \sigma^2 = \overline{x_0^2} \qquad (15\text{-}81)$$

In some problems it is of interest to know the probability that x_0 exceeds

---

[1] J. H. Laning, Jr., and R. H. Battin, "Random Processes in Automatic Control," p. 156, McGraw-Hill Book Company, Inc., New York, 1956.

OF AN ERGODIC RANDOM PROCESS

these are known

$\phi(\lambda)$	$p(x,y)$ $x = x(t)$ $y = x(t + \tau)$	$R(\tau)$	$\Phi(\omega)$	
No	No	No	No	$x(t)$
$\dfrac{1}{2\pi}\displaystyle\int_{-\infty}^{\infty} e^{-i\lambda x}\phi(\lambda)\,d\lambda$	$\displaystyle\int_{-\infty}^{\infty} p(x,y)\,dy$	No	No	$p(x)$
$-j\phi'(0)$	From $p(x)$	$[\overline{R(\tau)}]^{\frac12}$ See Note 2	$\left[\displaystyle\lim_{\epsilon\to 0}\int_{-\epsilon}^{\epsilon}\Phi(\omega)\,df\right]^{\frac12}$ See Note 3	$\bar{x}$
$-\phi''(0)$	From $p(x)$	$R(0)$	$\displaystyle\int_{-\infty}^{\infty}\Phi(\omega)\,df$	$\overline{x^2}$
—	From $p(x)$	No	No	$\phi(\lambda)$
No	—	No	No	$p(x,y)$ $x = x(t)$ $y = x(t + \tau)$
No	$\displaystyle\int_{-\infty}^{\infty} dx \int_{-\infty}^{\infty} dy\; xy\,p(x,y)$	—	$\mathcal{F}^{-1}[\Phi(\omega)]$	$R(\tau)$
No	From $R(\tau)$	$\mathcal{F}[R(\tau)]$	—	$\Phi(\omega)$

Note 2: The average is with respect to τ.   The sign of x cannot be determined.
Note 3: The sign of x cannot be determined.

some value a volts.   This is

$$Pr\,[x_0 \text{ exceeds } a] = 1 - \int_{-a}^{a} p(x_0)\,dx_0$$

$$= 1 - \frac{1}{(2\pi)^{\frac12}\sigma}\int_{-a}^{a} e^{-(x_0{}^2/2\sigma^2)}\,dx_0 \qquad (15\text{-}82)$$

Integrals like (15-82) are tabulated as functions of the limits in statistical tables.

Example 4.   The spectral densities of physical processes have a variety of shapes.   In fact, white noise passed through a filter of transfer function $Y(s)$ has a spectral density proportional to $|Y(j\omega)|^2$.

Noise occurring in transistors has a spectral density which varies as $1/f$ in a finite band of frequencies not including the origin.[1]

Experimental evidence indicates that the spectral density of vertical gust velocity in atmospheric turbulence varies inversely as the frequency squared.[2]   Since the gust phenomenon is characteristic of the atmos-

[1] W. R. Bennett, *op. cit.*
[2] H. Press and J. C. Houbalt, Some Applications of Generalized Harmonic Analysis to Gust Loads on Airplanes, *J. Aeronaut. Sci.*, **22**:17–26 (January, 1955).

phere, the spectral density will be a function of the velocity of the vehicle flying through it. For this reason the data are normalized, the variables used having dimensions of (ft/sec)² per radian/ft for spectral density and radians per foot for normalized frequency. The normalized variables are related to the ones used in this chapter as follows:

$$\Phi(\omega) = \frac{2\pi}{v} \Phi_1(\Omega) \tag{15-83}$$

$$\omega = v\Omega$$

where $\Phi(\omega)$ = spectral density, (ft/sec)² per cps
$\quad \Phi_1(\Omega)$ = spectral density, (ft/sec)² per radian/ft
$\quad \omega$ = frequency, radians/sec
$\quad \Omega$ = normalized frequency, radians/ft
$\quad v$ = vehicle velocity

The mean square gust velocity is

$$\overline{v_g^2} = \int_{-\infty}^{\infty} \Phi(\omega) \, df = \int_{-\infty}^{\infty} \Phi_1(\Omega) \, d\Omega \tag{15-84}$$

15-8. Summary of Methods of Description

The relationships among the various descriptions of an ergodic random process are shown in Table 15-1.

PROBLEMS

15-1. Find the spectral density from the correlation function if

$$R(\tau) = \frac{e^{-a|\tau|}}{4a^2} \left(|\tau| + \frac{1}{a} \right)$$

15-2. What filter would produce the output correlation function of Prob. 15-1 from a white-noise input?

15-3. How is $\Phi(0)$ related to $R(\tau)$?

15-4. Find the spectral density of the outputs of the filters shown if the input noise has spectral density

$$\Phi_i(\omega) = \frac{N_0}{1 + \omega^2}$$

What is the average output power in each case?

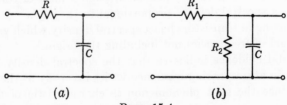

(a)   (b)

PROB. 15-4

15-5. Find by direct calculation the spectral density of the random-phase sine wave with uniformly distributed phase angle discussed in Sec. 14-5, Eq. (14-74).

15-6. Discuss the sudden application of white noise to an integrator.

15-7. Two independent stationary signals are present at the input of a system with transfer function $Y(s)$. What is the spectral density of the output in terms of the input spectra?

15-8. Referring to (15-51), start by writing

$$|A_T|^2 = \int_{-T}^{T} x(t_1)e^{-j\omega t_1}\, dt_1 \int_{-T}^{T} x(t_2)e^{j\omega t_2}\, dt_2$$

and show that

$$\widetilde{|A_T|^2} = \int_{-2T}^{2T} (2T - |\tau|)R(\tau)e^{-j\omega\tau}\, d\tau$$

Find at least one sufficient condition on $R(\tau)$ for the validity of (15-51).

15-9. Show that the approximating function in (15-57) has the necessary unit area.

15-10. Verify Eq. (15-60).

CHAPTER 16

DIFFERENCE EQUATIONS AND z TRANSFORMS

In this chapter we shall discuss linear difference equations and show how these can be solved by a transform method.

16-1. Difference Equations

To see how a difference equation can be used to describe a physical problem let us consider the network shown in Fig. 16-1.

We shall assume that all resistances except R_L on the end are of the same value R. Suppose that it is required to find the current in the nth

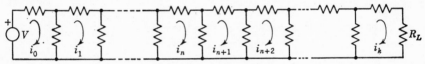

Fig. 16-1. Ladder network.

loop i_n. We could set up loop equations—there would be $k + 1$ of them, one for each loop—and solve for i_n. The equation for the $(n + 1)$st loop is

$$-Ri_n + 3Ri_{n+1} - Ri_{n+2} = 0 \qquad (16\text{-}1)$$

Instead of writing down the other k equations, we make the following observations:

1. Equation (16-1) is true for any n except -1 and $k - 1$, since the network is a repetitive structure and all loops (except the end ones) are alike.

2. Equation (16-1), together with end conditions, is sufficient to describe the network.

Let us rewrite (16-1) in a form to bring out the functional dependence on n:

$$i(n) - 3i(n + 1) + i(n + 2) = 0 \qquad (16\text{-}2)$$

This is a linear, constant-coefficient difference equation of the second order. Next, we shall develop a transform which will enable us to solve equations of this type. We shall return to this problem in Sec. 16-5.

246

Before going on, it should be made clear that there is a classical method for solution of equations like (16-2), just as there is a classical method for differential equations.[1] For a homogeneous equation like (16-2) one simply assumes a solution of the form

$$i(n) = a^n \tag{16-3}$$

This is substituted, a is determined, and arbitrary constants are evaluated from boundary conditions. When the right-hand side of the equation is not zero, the problem becomes more difficult but can still be solved.

The transform method has the same advantages for difference equations as for differential equations. Initial conditions (at $n = 0$) are introduced automatically; the solution can be found from tables; etc.

As in the case of differential equations, the transform method applies to linear equations. *A linear difference equation contains the dependent variable in the first degree only.* In (16-2) the addition of a term like $i(n)i(n + 1)$ would make the equation nonlinear.

In the next section, the Z transform will be developed.

16-2. Z Transforms

In the theory of sampled data systems[2] a sampled function is approximated by a train of impulse functions, each having an area equal to the function at the sampling instant. We define the sampled function corresponding to $f(t)$ as

$$f^*(t) = f(0)\delta_+(t) + f(T)\delta_+(t - T) + f(2T)\delta_+(t - 2T) + \cdots$$

$$= \sum_{n=0}^{\infty} f(nT)\delta_+(t - nT) \tag{16-4}$$

The sampling interval is T sec. The Laplace transform of the sampled function is

$$F^*(s) \triangleq \mathcal{L}[f^*(t)] = \sum_{n=0}^{\infty} f(nT)e^{-snT} \tag{16-5}$$

Evidently the variable in $F^*(s)$ is exp $(-snT)$ rather than s. It is convenient to make a change of variable

$$z = e^{sT} \tag{16-6}$$

---

[1] For a general discussion of difference equations and their classical solution, see L. M. Milne-Thomson, "The Calculus of Finite Differences," Macmillan & Co., Ltd., London, 1951.

[2] J. G. Truxal, "Automatic Feedback Control System Synthesis," chap. 9, McGraw-Hill Book Company, Inc., 1955.

so that (16-5) becomes

$$F_1(z) \triangleq \{\mathcal{L}[f^*(t)]\}_{e^{sT}=z} = \sum_{n=0}^{\infty} f(nT)z^{-n} \qquad (16\text{-}7)$$

Equation (16-7) can now be taken as the definition of a new transform, since it replaces a function of n by a new function of z. For our purposes it is convenient to set $T = 1$. We now define the Z transform:

$$Z[f(n)] \triangleq F(z) \triangleq \sum_{n=0}^{\infty} f(n)z^{-n} = f(0) + f(1)z^{-1} + f(2)z^{-2} + \cdots \qquad (16\text{-}8)$$

We now have a transform which, as we shall see, is useful in the solution of difference equations. We have shown how it evolves from the theory of sampled data systems, but this connection will not be pursued further here. We shall now investigate the properties of the transform defined by (16-8).

Let us first find the transforms of some functions of n.

We define the unit step function:

$$u(n) \triangleq \begin{cases} 1 & n \geq 0 \\ 0 & n < 0 \end{cases} \qquad (16\text{-}9)$$

so that from (16-8)

$$Z[u(n)] = \sum_{n=0}^{\infty} z^{-n} = 1 + z^{-1} + z^{-2} + \cdots = \frac{z}{z-1} \qquad (16\text{-}10)$$

as can be verified by long division. We thus have the pair

$$u(n) \qquad \Big| \qquad \frac{z}{z-1}$$

A table of pairs can be built up by expansion of rational fractions of z in much the same way that an integral table can be made by differentiating. For example,

$$\frac{z}{z-a} = 1 + az^{-1} + a^2z^{-2} + \cdots \qquad (16\text{-}11)$$

Comparing this with (16-8) we obtain the pair

$$a^n \qquad \Big| \qquad \frac{z}{z-a}$$

From this pair we obtain

$$e^{bn} \qquad \Big| \qquad \frac{z}{z-e^b}$$

Letting $b = j\omega_0$ we have

$$e^{j\omega_0 n} \quad \Bigg| \quad \frac{z}{z - e^{j w_0}}$$

If we write this pair as a sum of real and imaginary parts, we have

$$\cos \omega_0 n + j \sin \omega_0 n \quad \Bigg| \quad \frac{z(z - \cos \omega_0) + jz \sin \omega_0}{z^2 - 2z \cos \omega_0 + 1}$$

From the definition (16-8) we can write

$$\begin{aligned}
Z[f_1(n) + jf_2(n)] &= \sum_{n=0}^{\infty} [f_1(n) + jf_2(n)]z^{-n} \\
&= \sum_{n=0}^{\infty} f_1(n)z^{-n} + j \sum_{n=0}^{\infty} f_2(n)z^{-n} \\
&= F_1(n) + jF_2(n) \qquad (16\text{-}12)
\end{aligned}$$

so that the last pair can be written as two new pairs:

$$\cos \omega_0 n \quad \Bigg| \quad \frac{z(z - \cos \omega_0)}{z^2 - 2z \cos \omega_0 + 1}$$

$$\sin \omega_0 n \quad \Bigg| \quad \frac{z \sin \omega_0}{z^2 - 2z \cos \omega_0 + 1}$$

In this way a table such as appears at the end of the chapter can be built up.

16-3. Inversion

From (16-8) we write again the definition of the Z transform

$$Z[f(n)] = \sum_{n=0}^{\infty} f(n)z^{-n} = F(z) \qquad (16\text{-}13)$$

The inverse operation can be written

$$Z^{-1}[F(z)] = f(n) \qquad (16\text{-}14)$$

It is shown in Appendix A that $f(n)$ can be obtained from a complex inversion formula [analogous to (2-5) for $\mathcal{L}$ transforms]:

$$f(n) = \frac{1}{2\pi j} \oint F(z)z^{n-1} \, dz \qquad (16\text{-}15)$$

The existence of the inverse process implies conditions on $f(n)$ for uniqueness. We require that $f(n)$ be zero for negative n, just as we require that $f(t)$ be zero for negative t when $\mathcal{L}$ transforms are used (Sec. 2-1).

Equation (16-13) shows a way to find an inverse Z transform which does not require the contour integral (16-15). We write

$$F(z) = f(0) + f(1)z^{-1} + f(2)z^{-2} + \cdots \tag{16-16}$$

Evidently an expansion of $F(z)$ into a sum of inverse powers of z will exhibit $f(n)$ as coefficients of the expansion. When $F(z)$ is a rational fraction, the expansion can be made by long division. For example, let us find the inverse of the transform derived from (16-11):

$$Z^{-1}\left[\frac{z}{z - e^b}\right]$$

The long division process is shown below:

$$
\begin{array}{ll}
z & \underline{|z - e^b} \\
\underline{z - e^b} & \quad 1 + e^b z^{-1} + e^{2b}z^{-2} + \cdots \\
\quad e^b & \\
\quad \underline{e^b - e^{2b}z^{-1}} & \\
\qquad e^{2b}z^{-1} & \\
\qquad \underline{e^{2b}z^{-1} - e^{3b}z^{-2}} &
\end{array}
$$

We have now

$$\frac{z}{z - e^b} = 1 + e^b z^{-1} + e^{2b}z^{-2} + \cdots \tag{16-17}$$

and from (16-16) the values of $f(n)$ are

n	$f(n)$
0	1
1	e^b
2	e^{2b}

. . . .

In this case it is apparent that $f(n)$ is exp (bn). In general, however, such identification will not be possible, and (16-16) becomes a numerical method of inversion.

16-4. Properties of Z Transforms

Let us now investigate Z transforms of operations. The results are tabulated at the end of the chapter.

1. *Shifting.* First we shall find $Z[f(n + 1)]$. From (16-8)

$$Z[f(n + 1)] = \sum_{n=0}^{\infty} f(n + 1)z^{-n} = z \sum_{n=0}^{\infty} f(n + 1)z^{-(n+1)} \tag{16-18}$$

We can make a change in the variable of summation (the index) analogous to a change in variable of integration:

$$k = n + 1 \tag{16-19}$$

so that (16-18) becomes

$$\begin{aligned}
Z[f(n+1)] &= z \sum_{k=1}^{\infty} f(k)z^{-k} \\
&= z[f(1)z^{-1} + f(2)z^{-2} + \cdots] \tag{16-20}
\end{aligned}$$

We can make the sum start at $k = 0$ by adding and subtracting the missing term:

$$\begin{aligned}
Z[f(n+1)] &= z[-f(0) + f(0) + f(1)z^{-1} + f(2)z^{-2} + \cdots] \\
&= -zf(0) + z \sum_{k=0}^{\infty} f(k)z^{-k} = zF(z) - zf(0) \tag{16-21}
\end{aligned}$$

The operation-pair is

$$f(n+1) \quad | \quad zF(z) - zf(0)$$

In similar fashion the shifting can be extended to give the pairs

$$
\begin{array}{l|l}
f(n+2) & z^2F(z) - z^2f(0) - zf(1) \\
f(n+3) & z^3F(z) - z^3f(0) - z^2f(1) - zf(2) \\
f(n+a) \quad a > 0 & z^aF(z) - z^af(0) - z^{a-1}f(1) - \cdots - zf(a-1)
\end{array}
$$

We note two points about these pairs:

a. The shifting operation is transformed into an algebraic operation in z.

b. Initial conditions are introduced.

We therefore expect the Z transform to be effective in the solution of equations containing differences.

When the shift is in the opposite direction, we have

$$\begin{aligned}
Z[f(n-a)u(n-a)] &= \sum_{n=0}^{\infty} f(n-a)u(n-a)z^{-n} \\
&= z^{-a} \sum_{n=0}^{\infty} f(n-a)u(n-a)z^{-(n-a)} \tag{16-22}
\end{aligned}$$

A change of variable $k = n - a$ gives

$$\begin{aligned}
Z[f(n-a)u(n-a)] &= z^{-a} \sum_{k=-a}^{\infty} f(k)u(k)z^{-k} \\
&= z^{-a} \sum_{k=0}^{\infty} f(k)z^{-k} = z^{-a}F(z) \tag{16-23}
\end{aligned}$$

The unit step $u(k)$ makes all terms for negative k zero as required for uniqueness.

2. *Multiplication by n.* We write

$$Z[nf(n)] = \sum_{n=0}^{\infty} nf(n)z^{-n} = -z \sum_{n=0}^{\infty} f(n)[-nz^{-n-1}]$$

The term in the brackets is a derivative with respect to z:

$$Z[nf(n)] = -z \sum_{n=0}^{\infty} f(n) \frac{d}{dz} z^{-n}$$

$$= -z \frac{d}{dz} \sum_{n=0}^{\infty} f(n)z^{-n} = -z \frac{d}{dz} F(z) \qquad (16\text{-}24)$$

The result is

$$nf(n) \qquad \Big| \qquad -z \frac{d}{dz} F(z)$$

3. *Scale Change.* Let us find

$$Z^{-1}\left[F\left(\frac{z}{a}\right) \right] = Z^{-1}\left[\sum_{n=0}^{\infty} f(n) \left(\frac{z}{a}\right)^{-n} \right]$$

$$= Z^{-1}\left[\sum_{n=0}^{\infty} a^n f(n)z^{-n} \right]$$

$$= Z^{-1}\{Z[a^n f(n)]\} = a^n f(n) \qquad (16\text{-}25)$$

The pair is

$$a^n f(n) \qquad \Big| \qquad F\left(\frac{z}{a}\right)$$

4. *Summation.* The process corresponding to integration is summation. Let us find

$$Z\left[\sum_{k=0}^{n} f(k) \right]$$

First let us define

$$\sum_{k=0}^{n} f(k) \triangleq g(n) \qquad (16\text{-}26)$$

We can write down a relation between successive values of the sum:

$$g(n) = g(n-1)u(n-1) + f(n) \qquad (16\text{-}27)$$

This difference equation can be transformed using (16-23). We define

$$G(z) \triangleq Z[g(n)]$$
$$F(z) \triangleq Z[f(n)]$$

Then (16-27) becomes

$$G(z) = z^{-1}G(z) + F(z) \tag{16-28}$$

Solving for $G(z)$ we have

$$G(z) = \frac{F(z)}{1 - z^{-1}} = \frac{z}{z - 1} F(z) \tag{16-29}$$

from which we have the pair

$$\sum_{k=0}^{n} f(k) \qquad \Bigg| \qquad \frac{z}{z - 1} F(z)$$

5. Convolution. Since it is easier to deal with integrals than with sums, we use the inversion formula (16-15) in a formal way to find a convolution theorem for Z transforms.

We wish to find

$$Z^{-1}[F_1(z)F_2(z)]$$

From (16-15) we write

$$Z^{-1}[F_1(z)F_2(z)] = \frac{1}{2\pi j} \oint F_1(z)F_2(z)z^{n-1} \, dz \tag{16-30}$$

Substituting for $F_2(z)$ and then changing order of integration and summation, we have

$$Z^{-1}[F_1(z)F_2(z)] = \frac{1}{2\pi j} \oint dz \, F_1(z)z^{n-1} \sum_{k=0}^{\infty} f_2(k)z^{-k}$$

$$= \sum_{k=0}^{\infty} f_2(k) \frac{1}{2\pi j} \oint dz \, F_1(z)z^{n-1-k} \tag{16-31}$$

We now recognize the integral as $f_1(n - k)$ and write

$$Z^{-1}[F_1(z)F_2(z)] = \sum_{k=0}^{\infty} f_2(k)f_1(n - k) \tag{16-32}$$

6. Initial and Final Values. From the definition (16-8)

$$F(z) = \sum_{n=0}^{\infty} f(n)z^{-n} = f(0) + \frac{f(1)}{z} + \frac{f(2)}{z^2} + \cdots \tag{16-33}$$

We see that the initial value is obtained by letting $z \to \infty$:

$$f(0) = \lim_{z \to \infty} F(z) \tag{16-34}$$

The final value is not obtained so easily. Let us follow the method used for $\mathcal{L}$ transforms in Sec. 8-10. First we write

$$Z[f(n + 1) - f(n)] = \lim_{n \to \infty} \sum_{k=0}^{n} [f(k + 1) - f(k)]z^{-k} \quad (16\text{-}35)$$

From (16-21) we write the transform of the left-hand side:

$$zF(z) - zf(0) - F(z) = \lim_{n \to \infty} \sum_{k=0}^{n} [f(k + 1) - f(k)]z^{-k} \quad (16\text{-}36)$$

We now let $z \to 1$ in both sides of (16-36):

$$\begin{aligned}
\lim_{z \to 1} (z - 1)F(z) - f(0) &= \lim_{n \to \infty} \sum_{k=0}^{n} [f(k + 1) - f(k)] \\
&= \lim_{n \to \infty} \{[f(1) - f(0)] + [f(2) - f(1)] + \cdots \\
&\qquad + [f(n) - f(n - 1)] + [f(n + 1) - f(n)]\} \\
&= \lim_{n \to \infty} [-f(0) + f(n + 1)] = -f(0) + f(\infty)
\end{aligned}$$
$$(16\text{-}37)$$

so that we have finally, when the limit exists,

$$f(\infty) = \lim_{z \to 1} (z - 1)F(z) \quad (16\text{-}38)$$

16-5. Solution of Difference Equations

Having derived function and operation pairs, we proceed to solve difference equations by the Z-transform method.

Example 1: First-order Difference Equation. Let us solve the equation

$$f(n + 1) + 3f(n) = n \qquad f(0) = 1 \quad (16\text{-}39)$$

Transforming, with $Z[f(n)] \triangleq F(z)$ (and referring to the table in Sec. 16-8), we have

$$zF(z) - zf(0) + 3F(z) = \frac{z}{(z - 1)^2} \quad (16\text{-}40)$$

Now we solve for $F(z)$:

$$F(z) = \frac{z}{z + 3} + \frac{z}{(z + 3)(z - 1)^2} \quad (16\text{-}41)$$

The second term on the right can be expanded in partial fractions by the methods of Chap. 7. We shall expand

$$\frac{1}{(z + 3)(z - 1)^2} = \frac{A}{z + 3} + \frac{B}{(z - 1)^2} + \frac{C}{z - 1} \quad (16\text{-}42)$$

and then multiply both sides by z to put the terms in the form of entries in the table at the end of the chapter.   Applying the method of Sec. 7-2, we have

$$A = \tfrac{1}{16}$$
$$B = \tfrac{1}{4}$$
$$C = \left[\frac{d}{dz} \frac{1}{z+3} \right]_{z=1} = -\frac{1}{16} \qquad (16\text{-}43)$$

and (16-41) becomes

$$F(z) = \frac{z}{z+3} + \frac{1}{16} \frac{z}{z+3} + \frac{1}{4} \frac{z}{(z-1)^2} - \frac{1}{16} \frac{z}{z-1} \qquad (16\text{-}44)$$

from which

$$f(n) = \tfrac{1}{16}[17(-3)^n + 4n - 1] \qquad (16\text{-}45)$$

Example 2: Resistive Ladder Network.   The second-order difference equation for the network of Fig. 16-1 was given by (16-2) as

$$i(n) - 3i(n+1) + i(n+2) = 0 \qquad (16\text{-}46)$$

Transforming with $Z[i(n)] \triangleq I(z)$,

$$I(z) - 3zI(z) + 3zi(0) + z^2I(z) - z^2i(0) - zi(1) = 0 \quad (16\text{-}47)$$

Solving for $I(z)$, we have

$$I(z) = \frac{z[zi(0) - 3i(0) + i(1)]}{z^2 - 3z + 1} \qquad (16\text{-}48)$$

We can eliminate one of the initial conditions by writing the equation for the first loop of the network:

$$2Ri(0) - Ri(1) = V$$
$$i(1) = 2i(0) - \frac{V}{R} \qquad (16\text{-}49)$$

We now write for $I(z)$

$$I(z) = \frac{z[zi(0) - 3i(0) + 2i(0) - (V/R)]}{z^2 - 3z + 1}$$
$$= i(0) \frac{z(z - \{1 + [V/Ri(0)]\})}{z^2 - 3z + 1} \qquad (16\text{-}50)$$

Comparing with the pairs from the table

$\sinh \omega_0 n$	$\dfrac{z \sinh \omega_0}{z^2 - 2z \cosh \omega_0 + 1}$
$\cosh \omega_0 n$	$\dfrac{z(z - \cosh \omega_0)}{z^2 - 2z \cosh \omega_0 + 1}$

we rewrite (16-50) in a form for inversion:

$$I(z) = i(0) \frac{z(z - \tfrac{3}{2}) + z\{\tfrac{1}{2} - [V/Ri(0)]\}}{z^2 - 3z + 1} \qquad (16\text{-}51)$$

We identify cosh ω_0 by comparison with the transform pairs:

$$\cosh \omega_0 = \tfrac{3}{2}$$
$$\sinh \omega_0 = (\cosh^2 \omega_0 - 1)^{\frac{1}{2}} = \frac{\sqrt{5}}{2} \qquad (16\text{-}52)$$

Finally we have

$$i(n) = i(0) \left[\cosh \omega_0 n + \frac{\tfrac{1}{2} - [V/Ri(0)]}{\sqrt{5}/2} \sinh \omega_0 n \right] \qquad (16\text{-}53)$$

The value of $i(0)$ must now be found by substituting (16-53) into the equation for the end loop and solving for $i(0)$.

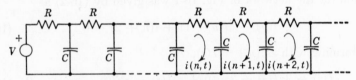

Fig. 16-2. Ladder network with capacitors.

Example 3 : Integro-difference Equation. Suppose the network of the preceding example is modified to include capacitors as shown in Fig. 16-2. The equation for the $(n + 1)$st loop is now, assuming initial conditions zero,

$$-\frac{1}{C} \int_0^t i(n,t)\, dt + \frac{2}{C} \int_0^t i(n + 1, t)\, dt + Ri(n + 1, t)$$
$$-\frac{1}{C} \int_0^t i(n + 2, t)\, dt = 0 \qquad (16\text{-}54)$$

The current is now a function of both loop index n and time t. If we $\mathcal{L}$-transform (16-54) with respect to t, with $\mathcal{L}[i(n,t)] \triangleq I(n,s)$, we have

$$-\frac{1}{Cs} I(n,s) + \frac{2}{Cs} I(n + 1, s) + RI(n + 1, s) - \frac{1}{Cs} I(n + 2, s) = 0$$
$$(16\text{-}55)$$

Next, we can Z-transform with $Z[I(n,s)] \triangleq \hat{I}(z,s)$:

$$-\frac{1}{Cs} \hat{I}(z,s) + \frac{2z}{Cs} \hat{I}(z,s) - \frac{2z}{Cs} I(0,s) + Rz\hat{I}(z,s) - RzI(0,s)$$
$$-\frac{z^2}{Cs} \hat{I}(z,s) + \frac{z^2}{Cs} I(0,s) + \frac{z}{Cs} I(1,s) = 0 \qquad (16\text{-}56)$$

from which

$$\hat{I}(z,s) = \frac{z[zI(0,s) - 2I(0,s) - RCsI(0,s) + I(1,s)]}{z^2 - (2 + RCs)z + 1} \quad (16\text{-}57)$$

This is in the same form as (16-48) and can be inverted in the same way to find $I(n,s)$. A second $\mathcal{L}$ inversion is needed to return to $i(n,t)$. The method of solution of problems like this follows the partial differential equation procedure of Chap. 11.

16-6. Poles and Zeros

The locations of poles and zeros of $F(z)$ in the z plane furnish information about the function of n corresponding to $F(z)$. Let us investigate the poles and zeros corresponding to a damped sinusoid with transform pair:

$$e^{-\alpha n} \sin \omega_0 n \quad \bigg| \quad \frac{ze^{-\alpha} \sin \omega_0}{z^2 - 2ze^{-\alpha} \cos \omega_0 + e^{-2\alpha}}$$

The transform has a zero at the z-plane origin. The poles are located at the roots of the denominator:

$$z = e^{-\alpha} \cos \omega_0 \pm \sqrt{e^{-2\alpha} \cos^2 \omega_0 - e^{-2\alpha}}$$
$$= e^{-\alpha}[\cos \omega_0 \pm j \sin \omega_0] = e^{-\alpha}e^{\pm j\omega_0} \quad (16\text{-}58)$$

The poles and zeros are shown in Fig. 16-3.

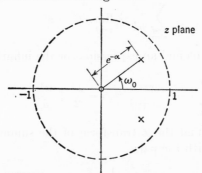

FIG. 16-3. z-plane poles and zeros for damped sinusoid.

Evidently with no damping ($\alpha = 0$) the roots lie on the unit circle. The unit circle is the boundary for stability just as the imaginary axis is the boundary in the s plane.

16-7. Summation of Infinite Series

In this section we shall discuss the application of Z transforms to the summation of infinite series. Starting with the pair

$$\sum_{k=0}^{n} f(k) \quad \bigg| \quad \frac{z}{z - 1} F(z)$$

we use the final-value theorem (16-38) and write

$$\sum_{k=0}^{\infty} f(k) = \lim_{z \to 1} (z - 1) \frac{z}{z - 1} F(z) = F(1) \qquad (16\text{-}59)$$

Example 1. Let us find the sum of

$$\sum_{n=0}^{\infty} \frac{x^n}{n!}$$

We start with the pair from Table 16-1,

$$\frac{1}{n!} \quad \Big| \quad e^{1/z}$$

Next from the operation pair (Table 16-2)

$$x^n f(n) \quad \Big| \quad F\left(\frac{z}{x}\right)$$

we form

$$\frac{x^n}{n!} \quad \Big| \quad e^{x/z}$$

Then

$$\sum_{n=0}^{\infty} \frac{x^n}{n!} = \lim_{z \to 1} e^{x/z} = e^x \qquad (16\text{-}60)$$

Example 2. Let us find an expression for the infinite series

$$\sum_{n=0}^{\infty} (-)^n \frac{x^{n+1}}{n + 1} = x - \frac{x^2}{2} + \frac{x^3}{3} - \cdots \qquad (16\text{-}61)$$

As before we shall find the z transform of the summand, and then let $z \to 1$. We start with the pair

$$x^n \quad \Big| \quad \frac{z}{z - x}$$

next

$$x^{n+1} \quad \Big| \quad \frac{xz}{z - x}$$

and

$$\frac{x^{n+1}}{n + 1} \quad \Big| \quad z \int_z^{\infty} \frac{xz}{z - x} \frac{dz}{z^2}$$

We now evaluate the integral

$$xz \int_z^{\infty} \frac{dz}{z(z - x)} = xz \left[\frac{1}{x} \ln \left(\frac{z - x}{z} \right) \right]_z^{\infty}$$

$$= -z \ln \left(\frac{z - x}{z} \right) \qquad (16\text{-}62)$$

The pair is now written

$$\frac{x^{n+1}}{n+1} \quad \bigg| \quad -z \ln\left(\frac{z-x}{z}\right)$$

Finally

$$(-)^n \frac{x^{n+1}}{n+1} \quad \bigg| \quad z \ln\left(\frac{z+x}{z}\right)$$

Now

$$\sum_{n=0}^{\infty} (-)^n \frac{x^{n+1}}{n+1} = \lim_{z\to 1} z \ln\left(\frac{z+x}{z}\right) = \ln(x+1) \qquad (16\text{-}63)$$

16-8. Transform Tables

In Tables 16-1 and 16-2 are tabulated Z-transform operation and function pairs.

TABLE 16-1. Z-TRANSFORM FUNCTION PAIRS

$f(n)$	$F(z)$
$u(n)$	$\dfrac{z}{z-1}$
n	$\dfrac{z}{(z-1)^2}$
n^2	$\dfrac{z(z+1)}{(z-1)^3}$
n^3	$\dfrac{z(z^2+4z+1)}{(z-1)^4}$
a^n	$\dfrac{z}{z-a}$
na^n	$\dfrac{az}{(z-a)^2}$
$\sin \omega_0 n$	$\dfrac{z \sin \omega_0}{z^2 - 2z \cos \omega_0 + 1}$
$\cos \omega_0 n$	$\dfrac{z(z - \cos \omega_0)}{z^2 - 2z \cos \omega_0 + 1}$
$\sinh \omega_0 n$	$\dfrac{z \sinh \omega_0}{z^2 - 2z \cosh \omega_0 + 1}$
$\cosh \omega_0 n$	$\dfrac{z(z - \cosh \omega_0)}{z^2 - 2z \cosh \omega_0 + 1}$
$e^{-\alpha n} \sin \omega_0 n$	$\dfrac{ze^{-\alpha} \sin \omega_0}{z^2 - 2ze^{-\alpha} \cos \omega_0 + e^{-2\alpha}}$
$e^{-\alpha n} \cos \omega_0 n$	$\dfrac{z(z - e^{-\alpha} \cos \omega_0)}{z^2 - 2ze^{-\alpha} \cos \omega_0 + e^{-2\alpha}}$
$\dfrac{1}{n!}$	$e^{1/z}$
$\dfrac{1}{(2n)!}$	$\cosh(z^{-1/2})$

TABLE 16-2. Z-TRANSFORM OPERATION PAIRS

$f(n)$	$F(z)$
$f(n+1)$	$zF(z) - zf(0)$
$f(n+2)$	$z^2F(z) - z^2f(0) - zf(1)$
$f(n+3)$	$z^3F(z) - z^3f(0) - z^2f(1) - zf(2)$
$f(n+a) \qquad a \geq 0$	$z^aF(z) - z^af(0) - \cdots - zf(a-1)$
$f(n-a)u(n-a) \qquad a \geq 0$	$z^{-a}F(z)$
$\Delta f(n) \triangleq f(n+1) - f(n)$	$(z-1)F(z) - zf(0)$
$nf(n)$	$-z \dfrac{d}{dz}F(z)$
$\dfrac{1}{n}f(n)$	$\displaystyle\int_z^\infty \dfrac{F(z)}{z}\,dz + \lim_{n\to 0}\dfrac{f(n)}{n}$
$\dfrac{1}{n+a}f(n)$	$\displaystyle z^a\int_z^\infty \dfrac{F(z)}{z^{a+1}}\,dz \qquad a>0$
$a^nf(n)$	$F\left(\dfrac{z}{a}\right)$
$\displaystyle\sum_{k=0}^n f(k)$	$\dfrac{z}{z-1}F(z)$
$\displaystyle\sum_{k=0}^n f_2(k)f_1(n-k)$	$F_1(z)F_2(z)$

$$f(0) = \lim_{z\to\infty} F(z)$$
$$f(\infty) = \lim_{z\to 1} (z-1)F(z)$$

PROBLEMS

16-1. Solve (16-2) by the classical method.

16-2. Obtain new pairs by expanding into series form and examining coefficients:

  a. $\dfrac{z}{(z+a)^2}$

  b. $\dfrac{z^2+z}{(z-1)^3}$

16-3. Use (16-32) to find

$$\mathcal{Z}^{-1}\left[\frac{z}{(z-1)^2}\right]$$

16-4. Show that

$$\mathcal{Z}\left[\frac{f(n)}{n}\right] = -\int \frac{F(z)}{z}\,dz$$

16-5. Show that

$$\mathcal{Z}[e^{-an}f(n)] = F(e^az)$$

16-6. Show from (16-53) that if the network is of infinite extent, the resistance at the input terminals is

$$R_0 = \frac{R(1 + \sqrt{5})}{2}$$

Hint: The coefficients of terms like e^{+n} must vanish.

16-7. Solve and check

    a. $y(n + 1) - 4y(n) = n^2$ $y(0) = 0$
    b. $y(n + 2) + 4y(n + 1) + 3y(n) = 0$ $y(0) = 1$ $y(1) = 1$
    c. $y(n + 1) - 5y(n) = \sin n$ $y(0) = 0$

16-8. Set up equations for the periodic structure shown.

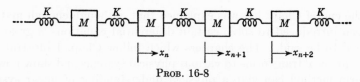

PROB. 16-8

16-9. Use Z transforms to find the sum of the geometric series

$$\sum_{k=0}^{n} a^k$$

16-10. Write a difference equation for

$$f(n) = \frac{1}{n!}$$

and solve it to show that

$$Z\left[\frac{1}{n!}\right] = e^{1/z}$$

16-11. Find the sums of the series below by Z transforms

    a. $\displaystyle\sum_{n=0}^{\infty} e^{-x(2n+1)}$ *Ans.* $\dfrac{1}{2 \sinh x}$

    b. $\displaystyle\sum_{n=0}^{\infty} \frac{\sin (n + 1)x}{n + 1}$ *Ans.* $\pm \dfrac{\pi}{2} - \dfrac{x}{2}$

    c. $\displaystyle\sum_{n=0}^{\infty} a^n \sin nx$ *Ans.* $\dfrac{a \sin x}{1 - 2a \cos x + a^2}$

16-12. Show that

$$Z\left[\left|\sin \frac{n\pi}{m}\right|\right] = \frac{z \sin \omega_0}{z^2 - 2z \cos \omega_0 + 1}\left[\frac{1 + z^{-m}}{1 - z^{-m}}\right] \qquad m = 1, 2, 3, \ldots$$

CHAPTER 17

INTRODUCTION TO MELLIN TRANSFORMS AND
A SUMMARY OF TRANSFORM METHODS

At the end of Chap. 1 it was explained that, although enough material had been introduced to solve certain differential equations, a great deal remained to be said. The chapters which follow Chap. 1 illustrate the application of $\mathcal{L}$ transforms to various physical systems and show how the transform method can give a physical understanding of linear systems through poles and zeros and through system functions. Partial differential equations were studied. Other transforms were introduced—Fourier and Z transforms. The transform method was applied to feedback systems and to statistical problems.

We can now make almost the same remark that appeared at the end of Chap. 1. There is still a great deal more that can be done with transforms. In this chapter we shall introduce another method—the Mellin transform—as an indication of further applications of the transform method. Section 17-5 contains a summary of transforms.

17-1. The Mellin Transform

The Mellin transform is defined:

$$\mathfrak{M}[f(r)] \triangleq \int_0^\infty f(r)r^{s-1}\,dr \triangleq F_M(s) \tag{17-1}$$

In order to use this transform to solve equations we are interested in knowing which operations performed on $f(r)$ transform into algebraic operations on $F_M(s)$. The derivative had a simple $\mathcal{L}$ transform. Let us find

$$\mathfrak{M}\left[\frac{df(r)}{dr}\right] = \int_0^\infty f'(r)r^{s-1}\,dr \tag{17-2}$$

Integrating by parts,

$$\mathfrak{M}\left[\frac{df(r)}{dr}\right] = f(r)r^{s-1}\Big|_0^\infty - \int_0^\infty f(r)(s-1)r^{s-2}\,dr \tag{17-3}$$

If s is suitably chosen and the transform exists, the first term on the right will vanish at both limits and we have

$$\mathfrak{M}\left[\frac{df(r)}{dr}\right] = -(s-1)\int_0^\infty f(r)r^{s-2}\,dr = -(s-1)F_M(s-1) \quad (17\text{-}4)$$

The shifting operation which yields $F_M(s-1)$ from $F_M(s)$ is not particularly simple.   Let us write this pair

$f(r)$	$F_M(s)$
$\dfrac{df(r)}{dr}$	$-(s-1)F_M(s-1)$

and go on with the search.

1. $r(d/dr)$ *Operator.*   We now find

$$\mathfrak{M}\left[r\frac{df(r)}{dr}\right] = \int_0^\infty \frac{df(r)}{dr}r^s\,dr = f(r)r^s\Big|_0^\infty - s\int_0^\infty f(r)r^{s-1}\,dr$$
$$= -sF_M(s) \quad (17\text{-}5)$$

This result is much simpler and indicates that the $\mathfrak{M}$ transform should be useful in solving differential equations with variable coefficients.

2. $r^2(d^2/dr^2)$ *Operator.*   Next we find

$$\mathfrak{M}\left[r^2\frac{d^2f(r)}{dr^2}\right] = \int_0^\infty \frac{d^2f}{dr^2}r^{s+1}\,dr$$
$$= \frac{df}{dr}r^{s+1}\Big|_0^\infty - \int_0^\infty \frac{df}{dr}(s+1)r^s\,dr$$
$$= -(s+1)f(r)r^s\Big|_0^\infty + s(s+1)\int_0^\infty f(r)r^{s-1}\,dr$$
$$= s(s+1)F_M(s) \quad (17\text{-}6)$$

where we have integrated by parts twice.   Again, this is a favorable result.   The pairs are:

$f(r)$	$F_M(s)$
$r\,\dfrac{df(r)}{dr}$	$-sF_M(s)$
$r^2\,\dfrac{d^2f(r)}{dr^2}$	$s(s+1)F_M(s)$

We observe that initial conditions are not introduced, so that it will be necessary to introduce these with impulse functions as discussed in Sec. 3-7.

17-2. Properties of $\mathfrak{M}$ Transforms

Before finding transforms of functions, let us derive some properties of the $\mathfrak{M}$ transform.

1. *Reciprocal.* Let us find

$$\mathfrak{M}\left[f\left(\frac{1}{r}\right)\right] = \int_0^\infty f\left(\frac{1}{r}\right) r^{s-1}\, dr \tag{17-7}$$

Making the substitution $x = 1/r$, we have

$$\mathfrak{M}\left[f\left(\frac{1}{r}\right)\right] = \int_\infty^0 f(x) x^{-s+1}(-x^{-2}\, dx)$$

$$= \int_0^\infty f(x) x^{-s-1}\, dx = F_M(-s) \tag{17-8}$$

2. *Multiplication by r^a.* Next we find

$$\mathfrak{M}[r^a f(r)] = \int_0^\infty f(r) r^{s+a-1}\, dr = F_M(s + a) \tag{17-9}$$

3. *Scale Factor.*

$$\mathfrak{M}[f(ar)] = \int_0^\infty f(ar) r^{s-1}\, dr \tag{17-10}$$

We make the change of variable $ar = x$:

$$\mathfrak{M}[f(ar)] = a^{-s+1} \int_0^\infty f(x) x^{s-1} \frac{1}{a}\, dx = a^{-s} F_M(s) \tag{17-11}$$

4. *Exponent.*

$$\mathfrak{M}[f(r^a)] = \int_0^\infty f(r^a) r^{s-1}\, dr \tag{17-12}$$

We make the change of variable $x = r^a$:

$$\mathfrak{M}[f(r^a)] = \begin{cases} \int_0^\infty f(x) x^{(1/a)(s-1)} \left(\frac{1}{a} x^{(1/a)-1}\, dx\right) & a > 0 \\ \int_\infty^0 f(x) x^{(1/a)(s-1)} \left(\frac{1}{a} x^{(1/a)-1}\, dx\right) & a < 0 \end{cases}$$

$$= \begin{cases} \dfrac{1}{a} \int_0^\infty f(x) x^{(s/a)-1}\, dx = \dfrac{1}{a} F_M\left(\dfrac{s}{a}\right) & a > 0 \\[2mm] -\dfrac{1}{a} \int_0^\infty f(x) x^{(s/a)-1}\, dx = -\dfrac{1}{a} F_M\left(\dfrac{s}{a}\right) & a < 0 \end{cases} \tag{17-13}$$

Table 17-1 is a summary of properties.

TABLE 17-1. $\mathfrak{M}$ TRANSFORMS OF OPERATIONS

$f(r)$	$F_M(s)$
$f'(r)$	$-(s-1)F_M(s-1)$
$rf'(r)$	$-sF_M(s)$
$r^2f''(r)$	$s(s+1)F_M(s)$
$f''(r)$	$(s-2)(s-1)F_M(s-2)$
$f\left(\dfrac{1}{r}\right)$	$F_M(-s)$
$r^a f(r)$	$F_M(s+a)$
$f(ar)$	$a^{-s}F_M(s)$
$f(r^a)$	$\begin{cases} \dfrac{1}{a}F_M\left(\dfrac{s}{a}\right) & a>0 \\[2mm] -\dfrac{1}{a}F_M\left(\dfrac{s}{a}\right) & a<0 \end{cases}$

17-3. $\mathfrak{M}$ Transforms of Functions

We start by transforming an impulse function

$$\mathfrak{M}[\delta(r-r_0)] = \int_0^\infty \delta(r-r_0)r^{s-1}\,dr = r_0^{s-1} \tag{17-14}$$

Next a step function:

$$\mathfrak{M}[u(r-r_0)] = \int_{r_0}^\infty r^{s-1}\,dr = \frac{r^s}{s}\Big|_{r_0}^\infty = -\frac{r_0^s}{s} \tag{17-15}$$

The transform of r^n is zero, but we can find

$$\mathfrak{M}[r^n u(r-r_0)] = \int_{r_0}^\infty r^{s+n-1}\,dr = \frac{r^{s+n}}{s+n}\Big|_{r_0}^\infty$$

$$= -\frac{r_0^{s+n}}{s+n} \tag{17-16}$$

Similarly,

$$\mathfrak{M}[r^n u(r_0-r)] = \frac{r_0^{s+n}}{s+n} \tag{17-17}$$

where we have changed the sign of the argument of the step function.
The transform of exp $(-ar)$ is a gamma function (see Sec. 2-4).

$$\mathfrak{M}[e^{-ar}] = \int_0^\infty e^{-ar}r^{s-1}\,dr = \frac{1}{a^s}\int_0^\infty e^{-ar}(ar)^{s-1}d(ar)$$

$$= a^{-s}\Gamma(s) \tag{17-18}$$

Table 17-2 is a summary of function pairs derived here.[1]

TABLE 17-2. $\mathfrak{M}$ TRANSFORMS OF FUNCTIONS

$f(r)$	$F_M(s)$
$\delta(r - r_0)$	$r_0{}^{s-1}$
$u(r - r_0)$	$-\dfrac{r_0{}^s}{s}$
$u(r_0 - r)$	$\dfrac{r_0{}^s}{s}$
$r^n u(r - r_0)$	$-\dfrac{r_0{}^{s+n}}{s + n}$
$r^n u(r_0 - r)$	$\dfrac{r_0{}^{s+n}}{s + n}$
e^{-ar}	$a^{-s}\Gamma(s)$

17-4. Solution of Equations by $\mathfrak{M}$ Transforms

From the pairs (17-5) and (17-6) we expect the $\mathfrak{M}$ transform to be applicable to equations of the form

$$A_n r^n \frac{d^n f(r)}{dr^n} + A_{n-1} r^{n-1} \frac{d^{n-1} f(r)}{dr^{n-1}} + \cdots + A_1 r \frac{df(r)}{dr} + A_0 f(r) = g(r)$$

$$(17\text{-}19)$$

This equation (sometimes called the *Euler equation*) can be solved by assuming a solution of the form r^p, substituting and solving for p, evaluating arbitrary constants, etc.[2] The transform method has the usual advantages over the classical method.

We now give examples of the solution of differential equations by $\mathfrak{M}$ transforms.

Example 1.   Let us solve

$$r \frac{dy}{dr} + 2y = r u(r - 1) \qquad (17\text{-}20a)$$

Transforming with $\mathfrak{M}[y(r)] \triangleq Y_M(s)$, we have

$$-s Y_M(s) + 2 Y_M(s) = -\frac{1}{s + 1}$$

$$Y_M(s) = \frac{1}{(s + 1)(s - 2)} = \frac{1}{3}\left[\frac{1}{s - 2} - \frac{1}{s + 1}\right] \qquad (17\text{-}20b)$$

From Table 17-2

$$y(r) = \tfrac{1}{3}(-r^{-2} + r)u(r - 1) \qquad (17\text{-}21)$$

[1] An extensive table of $\mathfrak{M}$ transforms will be found in A. Erdélyi, "Tables of Integral Transforms," vol. 1, McGraw-Hill Book Company, Inc., New York, 1954.

[2] See, for example, L. R. Ford, "Differential Equations," 2d ed., pp. 76, McGraw-Hill Book, Company, Inc., New York, 1955.

Note that since we have introduced no initial condition, the solution is zero at $r = 1$ where the step function in (17-20a) was applied.

Example 2.   The same equation with an initial condition is

$$r \frac{dy}{dr} + 2y = r \qquad y(r_0) = 1 \tag{17-22}$$

Following the procedure in Sec. 3-7 for introducing initial conditions, we write

$$y_1(r) \triangleq y(r)u(r - r_0)$$
$$r \frac{dy_1}{dr} = r \frac{dy}{dr} u(r - r_0) + r_0 y(r_0)\delta(r - r_0) \tag{17-23}$$

From (17-23) we write

$$r \frac{dy_1}{dr} + 2y_1 = \left[r \frac{dy}{dr} + 2y \right] u(r - r_0) + r_0 y(r_0)\delta(r - r_0) \tag{17-24}$$

The term in the brackets is, by (17-22), equal to r.   We now solve

$$r \frac{dy_1}{dr} + 2y_1 = ru(r - r_0) + r_0 y(r_0)\delta(r - r_0)$$
$$= ru(r - r_0) + r_0\delta(r - r_0) \tag{17-25}$$

Transforming as before, we have

$$-sY_M(s) + 2Y_M(s) = -\frac{r_0{}^{s+1}}{s + 1} + r_0{}^s$$

$$\begin{aligned}
Y_M(s) &= \frac{r_0{}^{s+1}}{(s - 2)(s + 1)} - \frac{r_0{}^s}{s - 2} \\
&= \frac{r_0{}^{s+1}}{3} \left[\frac{1}{s - 2} - \frac{1}{s + 1} \right] - \frac{r_0{}^s}{s - 2} \\
&= \frac{1}{3} \left[r_0{}^3 \frac{r_0{}^{s-2}}{s - 2} - \frac{r_0{}^{s+1}}{s + 1} \right] - r_0{}^2 \frac{r_0{}^{s-2}}{s - 2} \tag{17-26}
\end{aligned}$$

From Table 17-2,

$$\begin{aligned}
y(r) &= \left(-\frac{1}{3} \frac{r_0{}^3}{r^2} + \frac{1}{3} r + \frac{r_0{}^2}{r^2} \right) u(r - r_0) \\
&= \frac{1}{r^2} \left(r_0{}^2 - \frac{r_0{}^3}{3} \right) + \frac{1}{3} r \qquad r \geq r_0
\end{aligned} \tag{17-27}$$

Example 3: Poisson's Equation.   One way that equations of the type which can be solved by $\mathfrak{M}$ transforms arise is as follows.   Poisson's equation, which describes, for example, the potential due to a distribution of charge, is written in cylindrical coordinates[1]

$$\frac{1}{r} \frac{\partial}{\partial r} \left(r \frac{\partial \phi}{\partial r} \right) + \frac{1}{r^2} \frac{\partial^2 \phi}{\partial \theta^2} + \frac{\partial^2 \phi}{\partial z^2} = -\frac{1}{\epsilon} \rho(r,\theta,z) \tag{17-28}$$

[1] See, for example, J. A. Stratton, "Electromagnetic Theory," p. 51, 162, McGraw-Hill Book Company, Inc., New York, 1941.

where ϵ is the inductive capacity of the medium and ρ is charge density. If we assume no variations of ϕ with z, we can write (17-28) in the form

$$r^2 \frac{\partial^2 \phi(r,\theta)}{\partial r^2} + r \frac{\partial \phi(r,\theta)}{\partial r} + \frac{\partial^2 \phi(r,\theta)}{\partial \theta^2} = r^2 f(r,\theta) \qquad (17\text{-}29)$$

$\mathcal{L}$-transforming with respect to θ, we have, with $\mathcal{L}[\phi(r,\theta)] \triangleq \Phi(r,s)$, etc.,

$$r^2 \frac{d^2 \Phi(r,s)}{dr^2} + r \frac{d\Phi(r,s)}{dr} + s^2 \Phi(r,s) = r^2 F(r,s) + s\phi(r,0) + \frac{\partial \phi(r,0)}{\partial \theta} \qquad (17\text{-}30)$$

The equation remaining can be $\mathfrak{M}$-transformed.

TABLE 17-3. SUMMARY OF TRANSFORMS

Operation	Transform	Transform of operation	Reference
$\dfrac{df(t)}{dt}$	Laplace transform $F(s) = \int_0^\infty e^{-st} f(t)\,dt$	$sF(s) - f(0)$	Chap. 1
	Fourier transform $F(\omega) = \int_{-\infty}^\infty e^{-i\omega t} f(t)\,dt$	$j\omega F(\omega)$	Chap. 13
	Finite Fourier transform $f_s(n) = \int_0^L f(t) \sin \omega_n t\,dt$ $\hat{f}_c(n) = \int_0^L f(t) \cos \omega_n t\,dt$	$-\omega_n \hat{f}_c(n)$ $(-)^n y(L) - y(0) + \omega_n \hat{f}_s(n)$	Chap. 12
$f(n+1)$	Z transform $F(z) = \sum_{n=0}^\infty f(n) z^{-n}$	$zF(z) - zf(0)$	Chap. 16
$t\,\dfrac{df(t)}{dt}$	Mellin transform $F_M(s) = \int_0^\infty f(t) t^{s-1}\,dt$	$-sF_M(s)$	Chap. 17
$\dfrac{1}{t} \dfrac{d}{dt}\left(t\,\dfrac{df(t)}{dt} \right) - \dfrac{\nu^2}{t^2} f(t)$	Hankel transform $F(s) = \int_0^\infty f(t) t J_\nu(st)\,dt$	$-s^2 F(s)$	Sneddon[1]
$\dfrac{d^2 f(t)}{dt^2} + \dfrac{b_0}{t} \dfrac{df(t)}{dt}$	Meijer transform $F_\nu(s) = \int_0^\infty f(t) t^{\nu+1} K_\nu(st)\,dt$ (K_ν is a modified Bessel function)	$s^2 F_\nu(s) - \dfrac{\Gamma(\nu+1)2^\nu}{s^\nu} f(0)$ $\nu = \dfrac{b_0 - 1}{2}$	Aseltine[2] Erdélyi[3]
$\dfrac{d}{dt}\left[(1-t^2)\,\dfrac{df(t)}{dt} \right]$	Legendre transform $F(n) = \int_{-1}^1 f(t) P_n(t)\,dt$ (P_n is a Legendre polynomial)	$-n(n+1)F(n)$	Tranter[4]

[1] I. N. Sneddon, "Fourier Transforms," sec. 10, McGraw-Hill Book Company, Inc., New York, 1951.

[2] J. A. Aseltine, A Transform Method for Linear Time-varying Systems, *J. Appl. Phys.*, **25**: 761–764 (June, 1954).

[3] A. Erdélyi, "Tables of Integral Transforms," vol. 2, chap. 10, McGraw-Hill Book Company, Inc., New York, 1954.

[4] C. J. Tranter, "Integral Transforms in Mathematical Physics," p. 96, John Wiley & Sons, Inc., New York, 1951.

17-5. Summary of Transform Methods

Transforms other than those introduced so far are used to solve physical problems, but most of these involve the higher transcendental functions (like Bessel functions in Hankel transforms) and are beyond the scope of this book.[1]   Some of these are tabulated in Table 17-3 along with the transforms we have used so far.   The emphasis in tabulation is on the operation which can be transformed most simply, since this is the key to solution of equations.

PROBLEMS

17-1. Show that

$a.$ $\mathfrak{M}\left[\left(x\dfrac{d}{dx}\right)^n f(x)\right] = (-)^n s^n F_M(s)$

$b.$ $\mathfrak{M}[f''(x)] = (s-2)(s-1)F_M(s-2)$

17-2. Show that $\mathfrak{M}[e^{-x^2}] = \frac{1}{2}\Gamma(s/2)$.

17-3. Solve and check

$a.$ $r\dfrac{dy}{dr} + y = 1 + r \qquad y(1) = 2$

$b.$ $r\dfrac{dy}{dr} + y = 4r^3 + 3r^2 \qquad y(1) = 3 \qquad\qquad Ans. \quad y = \dfrac{1}{r} + r^2 + r^3$

17-4. A transform is defined:

$$\mathfrak{M}_1[f(r)] \triangleq \int_a^\infty f(r)r^{s-1}\,dr$$

Find transforms of operations and functions, and discuss applications.   Compare with the $\mathfrak{M}$ transform.

[1] See Erdélyi, *op. cit.*, vols. 1 and 2, for tabulation of other transforms.   Also see E. C. Titchmarsh, "Introduction to the Theory of Fourier Integrals," chap. 8, Oxford University Press, London, 1948, for a discussion of general transforms.

COMPLEX VARIABLES

Although the material in the chapters of this book can be understood and used without it, it is worthwhile to discuss the theory of complex variables briefly. A more thorough discussion can be found in any of the many books available.[1] Here we shall present definitions and theorems leading to deeper understanding of transform topics.

A-1. Basic Definitions and Algebra

A complex variable is a pair of real numbers written

$$z = x + jy \tag{A-1}$$

We call x the real part and y the imaginary part of z:

$$\begin{aligned} x &= \mathrm{Re}\ [z] \\ y &= \mathrm{Im}\ [z] \end{aligned} \tag{A-2}$$

The j is a number such that $j^2 = -1$.

The ordinary rules of algebra hold for complex variables. For example, if

$$z_1 = x_1 + jy_1 \tag{A-3}$$
$$z_2 = x_2 + jy_2 \tag{A-4}$$

Then

$$\begin{aligned} z_1 \pm z_2 &= (x_1 \pm x_2) + j(y_1 \pm y_2) \\ z_1 z_2 &= (x_1 + jy_1)(x_2 + jy_2) = (x_1 x_2 - y_1 y_2) + j(x_1 y_2 + x_2 y_1) \end{aligned} \tag{A-5}$$

etc.

We can write z in polar form as follows:

$$z = |z|e^{j\theta} = |z|(\cos \theta + j \sin \theta) \tag{A-6}$$

where

$$\theta = \tan^{-1} \frac{y}{x} \tag{A-7}$$

and

$$|z| = (x^2 + y^2)^{\frac{1}{2}} \tag{A-8}$$

---

[1] For example, R. V. Churchill, "Introduction to Complex Variables and Applications," McGraw-Hill Book Company, Inc., New York, 1948. Also, E. T. Copson, "Theory of Functions of a Complex Variable," Oxford University Press, London, 1935.

The complex conjugate of z is defined as

$$z^* \triangleq x - jy \qquad \text{(A-9)}$$

From this definition it follows that

$$z + z^* = 2 \text{ Re } [z] \qquad \text{(A-10)}$$

and

$$z - z^* = 2j \text{ Im } [z] \qquad \text{(A-11)}$$

We often represent complex numbers as points in a complex plane. The imaginary part is the ordinate, and the real part is the abscissa. The plot is shown in Fig. A-1.

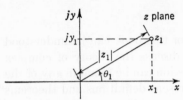

FIG. A-1. Complex z plane.

We can think of z as being a function of some real variable, say t. Then we write

$$z = z(t) \qquad \text{(A-12)}$$

If we let z_1 be the value of z at some time t_1, then

$$z_1 = z(t_1) = x(t_1) + jy(t_1) \qquad \text{(A-13)}$$

As t changes, a path will be traced out in the complex plane as in Fig. A-2. As an example, suppose

$$z(t) = 2 + jt \qquad \text{(A-14)}$$

We get the path shown in Fig. A-3 as t increases.

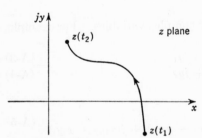

FIG. A-2. Path in the z plane.

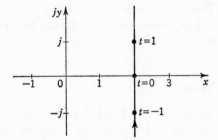

FIG. A-3. Path traced by $2 + jt$.

A-2. Analysis

By "analysis" is meant the calculus of complex variables. Derivatives and integrals of a function of a complex variable are defined in the same way as those for functions of a real variable—with one important difference: We must take into account *directions* in which the variable z changes in the complex plane. This was not necessary when dealing with functions of a real variable, since the real variable can change only its magnitude. The complex variable, on the other hand, can move about the complex plane.

The derivative of a function of a complex variable $f(z)$ is defined as

$$f'(z) = \frac{df(z)}{dz} \triangleq \lim_{\Delta z \to 0} \frac{f(z + \Delta z) - f(z)}{\Delta z} \tag{A-15}$$

Here $f(z)$ might be, for example, $\sin z$ or z^2. The change in z, Δz, can be thought of as the path in the z plane traced in moving from $z + \Delta z$ to z, as shown in Fig. A-4. There are infinitely many possibilities, and our definition of $f'(z)$ seems to be incomplete. A function of z might have one derivative when a point is approached horizontally and another when it is approached vertically. This is, indeed, often the case. However, there is a large class of functions for which the value of $f'(z)$ is independent of the direction of approach to the point z.

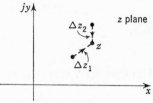

FIG. A-4. Increments in z.

A function $f(z)$ which is single valued and which possesses a unique derivative is called *analytic*. For example,

$$f(z) = \frac{1}{z} \tag{A-16}$$

is analytic everywhere except at $z = 0$. The derivative, which is taken in the usual way, is

$$f'(z) = -\frac{1}{z^2} \tag{A-17}$$

and at the origin this is not defined (it becomes infinite).

The function

$$f(z) = z^* \tag{A-18}$$

is not analytic anywhere because its derivative in the x direction

$$\frac{\partial}{\partial x}(z^*) = \frac{\partial}{\partial x}(x - jy) = 1 \tag{A-19}$$

is not equal to its derivative in the jy direction

$$\frac{\partial}{\partial(jy)}(z^*) = \frac{1}{j}\frac{\partial}{\partial y}(x - jy) = -1 \tag{A-20}$$

A point in the complex plane where $f(z)$ is *not* analytic is called a *singular point*, and $f(z)$ is said to have a *singularity* there.

Of principal interest to us is the kind of singularity known as a pole and discussed in Sec. 2-5. When we have a point z_1 such that $|f(z)| \to \infty$ as $z \to z_1$, and $(z - z_1)^n f(z)$ is finite and not zero there, then $f(z)$ has an nth-order pole at z_1.

A function can be tested very simply for analyticity by the Cauchy-Riemann equations.[1]   This test is both necessary and sufficient.

A function of z can be written

$$f(z) = u(x,y) + jv(x,y) \tag{A-21}$$

For example, if

$$f(z) = z^2 = (x + jy)^2 = (x^2 - y^2) + j(2xy) \tag{A-22}$$

then

$$\begin{aligned} u(x,y) &= x^2 - y^2 \\ v(x,y) &= 2xy \end{aligned} \tag{A-23}$$

The Cauchy-Riemann condition for analyticity requires that

$$\frac{\partial u}{\partial x} = \frac{\partial v}{\partial y} \tag{A-24}$$

and

$$\frac{\partial u}{\partial y} = -\frac{\partial v}{\partial x} \tag{A-25}$$

In the example above

$$\begin{array}{ll} \dfrac{\partial u}{\partial x} = 2x & \dfrac{\partial v}{\partial y} = 2x \\[2mm] \dfrac{\partial u}{\partial y} = -2y & \dfrac{\partial v}{\partial x} = 2y \end{array} \tag{A-26}$$

and we are assured that $f(z) = z^2$ is analytic for finite z.

A-3. Complex Integration

The integral of a function of a complex variable is defined as the limit of a sum, just as for functions of a real variable:

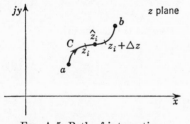

FIG. A-5. Path of integration.

$$\int_a^b f(z)\, dz = \lim_{\substack{n \to \infty \\ |\Delta z| \to 0}} \sum_{i=1}^{n} f(\hat{z}_i)\, \Delta z \tag{A-27}$$

where $\hat{z}_i$ is some value between z_i and $z_i + \Delta z$. In writing the integral we must specify the path of integration as in Fig. A-5.   Another way to write this integral is

$$\int_C f(z)\, dz \tag{A-28}$$

It should be emphasized that the function of z to be integrated and the path of integration can be selected independently.

[1] For a proof, see Churchill, *op. cit.*, pp. 28–30.

In the next section we shall state a theorem which provides the value of an integral over a closed path, or *contour*. Before this, however, it is instructive to evaluate such an integral by more elementary means.

Suppose we ask for the value of

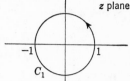

$$\int_{C_1} \frac{dz}{z} \qquad (A-29)$$

FIG. A-6. Path C_1.

where C_1 is the unit circle traced in the counter-clockwise direction in the z plane[1] as shown in Fig. A-6. When z is on C_1, we have

$$z = e^{j\theta} \qquad 0 \le \theta \le 2\pi \qquad (A-30)$$

and

$$dz = je^{j\theta}\, d\theta \qquad (A-31)$$

Substituting these in the integral, we have

$$\int_{C_1} \frac{dz}{z} = \int_0^{2\pi} \frac{je^{j\theta}\, d\theta}{e^{j\theta}} = j \int_0^{2\pi} d\theta = 2\pi j \qquad (A-32)$$

A-4. Cauchy's Residue Theorem

In preparation for the theorem mentioned in the previous section we must digress long enough to define the residue of a function.[2] A complete definition would require the development of the series expansion of a function $f(z)$ near a singularity. We shall state here only that if such an expansion of $f(z)$ about a singular point z_0 contains a term of the form $b_1/(z - z_0)$, then b_1 is called the *residue* of $f(z)$ at z_0.

The residue of $f(z)$ at an nth-order pole at $z = z_0$ is given by

$$b_1 = \frac{1}{(n-1)!} \left[\frac{d^{n-1}}{dz^{n-1}} (z - z_0)^n f(z) \right]_{z=z_0} \qquad (A-33)$$

For a first-order, or simple, pole we have

$$b_1 = [(z - z_0)f(z)]_{z=z_0} \qquad (A-34)$$

Also, when $f(z)$ can be written as a fraction

$$f(z) = \frac{n(z)}{d(z)} \qquad (A-35)$$

then we can find b_1 for a first-order pole at z_0 from

$$b_1 = \frac{n(z_0)}{d'(z_0)} \qquad (A-36)$$

---

[1] When the path is closed, the integral is often written $\oint$. The arrow indicates the direction in which the path is traversed.

[2] See Sec. 7-4.

As an example of the use of (A-33) suppose

$$f(z) = \frac{z^2}{(z-1)^2} \tag{A-37}$$

$f(z)$ has a second-order pole at $z_0 = 1$. Then

$$b_1 = \frac{1}{1!}\left[\frac{d}{dz}(z-1)^2 \frac{z^2}{(z-1)^2} \right]_{z=1} = [2z]_{z=1} = 2 \tag{A-38}$$

We now present one of the most important (and remarkable) theorems in analysis.

Cauchy's Theorem: If $f(z)$ is analytic on and inside a contour C, except at a finite number of singular points $z_1, z_2, \ldots$, within C, then

$$\oint_C f(z)\,dz = 2\pi j[k_1 + k_2 + \cdots + k_n + \cdots]$$

where k_n is the residue of $f(z)$ at z_n and C is taken in the counterclockwise sense.[1]

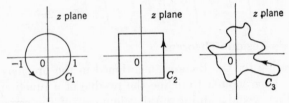

FIG. A-7. Integration contours.

To illustrate this theorem, let us return to a problem used earlier: to show that for the contour in Fig. A-6

$$\oint \frac{dz}{z} = 2\pi j \tag{A-39}$$

Here $f(z) = 1/z$. There is a simple pole at $z = 0$. Therefore

$$k = \left[z \cdot \frac{1}{z} \right]_{z=0} = 1 \tag{A-40}$$

and by Cauchy's theorem

$$\oint \frac{dz}{z} = 2\pi jk = 2\pi j \tag{A-41}$$

Notice that the contour C need not be specified other than that it be closed and include the origin. According to the theorem it is true that

$$\oint_{C_1} \frac{dz}{z} = \oint_{C_2} \frac{dz}{z} = \oint_{C_3} \frac{dz}{z} = 2\pi j \tag{A-42}$$

where C_1, C_2, and C_3 are given in Fig. A-7.

[1] The integral in the clockwise sense has the same value but with negative sign.

A-5. Contour Integral Representation of the Unit Step and Impulse Functions

As an application of the material in the preceding sections and in preparation for the discussion of the $\mathcal{L}^{-1}$ operation, we shall derive representations of the unit step and impulse functions.

The unit step funcion $u(t)$ is defined as

$$u(t) = \begin{cases} 0 & t < 0 \\ 1 & t > 0 \end{cases} \tag{A-43}$$

We shall now prove that $u(t)$ can be represented as the integral:

$$u(t) = \frac{1}{2\pi j} \int_{\sigma_1 - j\infty}^{\sigma_1 + j\infty} \frac{e^{st}}{s} \, ds \qquad \sigma_1 > 0 \tag{A-44}$$

The path of integration in the s plane is a straight line as shown in Fig A-8. Our proof will be based on Cauchy's theorem. First we must

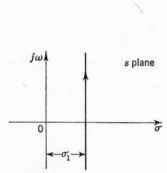

FIG. A-8. Path of integration for Eq. (A-44).

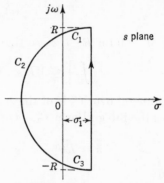

FIG. A-9. Closure of path of integration.

somehow close the path of integration. Suppose we redraw the path as in Fig. A-9 and then let $R \to \infty$. We now have more than the desired path. The complete contour integral can be written as the sum of the parts:

$$\oint = \int_{\sigma_1 - jR}^{\sigma_1 + jR} + \int_{C_1} + \int_{C_2} + \int_{C_3} \tag{A-45}$$

The first term on the right will be the integral we started with when $R \to \infty$. The others, we shall now show, tend to zero as $R \to \infty$. We shall then have the desired integral in terms of an integral over a closed path which can be evaluated by residues.

To show that an integral of a complex function $\to 0$, it is sufficient to show that its absolute value $\to 0$. This is true, since we can always write

$$f(z) = |f(z)|e^{j\theta} \tag{A-46}$$

so that if $|f(z)| \to 0$, then $f(z) \to 0$.

It can be shown also that[1]

$$\left| \int f(z) \, dz \right| \leq \int |f(z) \, dz| \tag{A-47}$$

and

$$\int |f(z) \, dz| \leq ML \tag{A-48}$$

where M is the maximum value of $|f(z)|$ and L is the length of the path of integration.

Keeping these things in mind, let us examine the integral

$$\int_{C_1} \frac{e^{st}}{s} \, ds$$

Now on C_1, if $R \to \infty$, $L = \sigma_1$. The integrand is

$$f(s) = \frac{e^{t\sigma} e^{jRt}}{\sigma + jR} \tag{A-49}$$

so that

$$M = |f(s)|_{\max} = \frac{e^{t\sigma_1}}{(\sigma_1^2 + R^2)^{1/2}} \tag{A-50}$$

Now $M \to 0$ for $R \to \infty$, so that for all finite values of t

$$\left| \int_{C_1} \right| \leq ML = M\sigma_1 \to 0 \qquad \text{as } R \to \infty \tag{A-51}$$

The integral over C_3 is of the same form as this and also vanishes. This leaves the integral over C_2. On C_2

$$s = Re^{j\theta} \qquad ds = jRe^{j\theta} \, d\theta \qquad \frac{\pi}{2} \leq \theta \leq \frac{3\pi}{2} \tag{A-52}$$

Then

$$\int_{C_2} \frac{e^{st}}{s} \, ds = \int_{\pi/2}^{3\pi/2} \frac{e^{tR \exp{(j\theta)}} jRe^{j\theta} \, d\theta}{Re^{j\theta}} = j \int_{\pi/2}^{3\pi/2} e^{tR \cos \theta} e^{jtR \sin \theta} \, d\theta \tag{A-53}$$

We now write from (A-47)

$$\left| j \int_{\pi/2}^{3\pi/2} e^{tR \cos \theta} e^{jtR \sin \theta} \, d\theta \right| \leq \int_{\pi/2}^{3\pi/2} e^{tR \cos \theta} \, d\theta = 2 \int_{\pi/2}^{\pi} e^{tR \cos \theta} \, d\theta$$

$$= 2 \int_0^{\pi/2} e^{tR \cos (\phi + \pi/2)} \, d\phi$$

$$= 2 \int_0^{\pi/2} e^{-tR \sin \phi} \, d\phi \tag{A-54}$$

The curve of $\sin \phi$ vs. ϕ for $0 \leq \phi \leq \pi/2$ is shown in Fig. A-10. The sine curve can be bounded by straight lines in the figure, so that we can write

$$\frac{2\phi}{\pi} \leq \sin \phi \leq \phi \qquad 0 \leq \phi \leq \frac{\pi}{2} \tag{A-55}$$

[1] See Churchill, op. cit., p. 79, for a proof.

This is known as *Jordan's inequality*. Returning now to (A-54) we see that for $t > 0$ we can write

$$2 \int_0^{\pi/2} e^{-tR \sin \phi} \, d\phi \leq 2 \int_0^{\pi/2} e^{-tR2\phi/\pi} \, d\phi = \frac{\pi}{tR} (1 - e^{-tR}) \quad \text{(A-56)}$$

which tends to zero as $R \to \infty$. Therefore the integral over C_2 vanishes,[1] and we have

$$\int_{\sigma_1 - j\infty}^{\sigma_1 + j\infty} \frac{e^{st}}{s} \, ds = \oint \frac{e^{st}}{s} \, ds \quad \text{(A-57)}$$

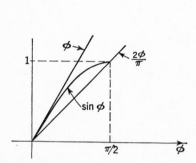

FIG. A-10. Bounds on $\sin \phi$ for $0 \leq \phi \leq \pi/2$.

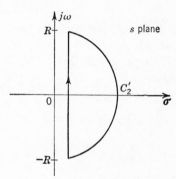

FIG. A-11. Closure for $t < 0$.

Now the residue of the integrand of $\oint$ at $s = 0$ is

$$k = \left[s \frac{e^{st}}{s} \right]_{s=0} = 1 \quad \text{(A-58)}$$

so that

$$\int_{\sigma_1 - j\infty}^{\sigma_1 + j\infty} \frac{e^{st}}{s} \, ds = 2\pi j \qquad t > 0 \qquad \sigma_1 > 0 \quad \text{(A-59)}$$

or

$$\frac{1}{2\pi j} \int_{\sigma_1 - j\infty}^{\sigma_1 + j\infty} \frac{e^{st} \, ds}{s} = 1 \qquad t > 0 \quad \text{(A-60)}$$

In the case when $t < 0$, the integral over C_2 does not vanish. We can, however, close the contour as in Fig. A-11, and now, by a similar argument, $\int_{C_2'} \to 0$ as $R \to \infty$ for $t < 0$. We then have over this new path

$$\oint \frac{e^{st}}{s} \, ds = \int_{\sigma_1 - j\infty}^{\sigma_1 + j\infty} \frac{e^{st}}{s} \, ds \quad \text{(A-61)}$$

[1] The integral over C_2 will always vanish for $t > 0$ if the integrand is $f(s)e^{st}$ and $f(s)$ is a proper rational fraction (*Jordan's Lemma*—cf. Copson, *op. cit.*, p. 137).

But the integrand of $\oint$ is analytic (has no poles) inside the contour. Hence there are no residues, and hence the integral is zero by Cauchy's theorem. We therefore have

$$\frac{1}{2\pi j} \int_{\sigma_1 - j\infty}^{\sigma_1 + j\infty} \frac{e^{st}}{s}\, ds = 0 \qquad t < 0 \qquad \sigma_1 > 0 \tag{A-62}$$

and our assertion (A-44) is true:

$$u(t) = \frac{1}{2\pi j} \int_{\sigma_1 - j\infty}^{\sigma_1 + j\infty} \frac{e^{st}}{s}\, ds \qquad \sigma_1 > 0 \tag{A-63}$$

This proof, although lengthy, serves to illustrate some of the techniques involved in contour integration.

The impulse function $\delta(t - \tau)$ is the derivative[1] of the step $u(t - \tau)$

$$\delta(t - \tau) = \frac{d}{dt} u(t - \tau) \tag{A-64}$$

We can write from (A-63)

$$u(t - \tau) = \frac{1}{2\pi j} \int_{\sigma_1 - j\infty}^{\sigma_1 + j\infty} \frac{e^{s(t-\tau)}}{s}\, ds \tag{A-65}$$

Taking the derivative with respect to t (which is not the variable of integration) we have formally

$$\delta(t - \tau) = \frac{d}{dt} u(t - \tau) = \frac{1}{2\pi j} \int_{\sigma_1 - j\infty}^{\sigma_1 + j\infty} e^{s(t-\tau)}\, ds \tag{A-66}$$

A-6. The Laplace Inversion Formula

We shall use the expression for $\delta(t - \tau)$ just derived together with the property of impulse functions (3-6):

$$\int_{\tau - \epsilon}^{\tau + \epsilon} f(t)\, \delta(t - \tau)\, dt = f(\tau) \qquad \epsilon > 0 \tag{A-67}$$

to derive the $\mathcal{L}$ inversion formula.   First, let us write the definition of $\mathcal{L}[f(t)]$:

$$F(s) \triangleq \mathcal{L}[f(t)] = \int_0^\infty f(t) e^{-st}\, dt \qquad s = \sigma + j\omega \tag{A-68}$$

For convergence we require that the real part of s be greater than some number σ_1:

$$\sigma > \sigma_1 \tag{A-69}$$

where σ_1 is such that the integral

$$\int_0^\infty |f(t)| e^{-\sigma_1 t}\, dt \qquad \text{exists} \tag{A-70}$$

---

[1] See Sec. 3-3.

Now let us multiply both sides of (A-68) by $e^{st}/2\pi j$ and integrate:

$$\frac{1}{2\pi j} \int_{\sigma_2-j\infty}^{\sigma_2+j\infty} F(s)e^{st}\,ds = \frac{1}{2\pi j} \int_{\sigma_2-j\infty}^{\sigma_2+j\infty} e^{st}\,ds \int_0^{\infty} f(\tau)e^{-s\tau}\,d\tau \qquad \text{(A-71)}$$

where $\sigma_2 > \sigma_1$ to ensure convergence. Interchanging the order of integration, we have (using A-66)

$$\frac{1}{2\pi j} \int_{\sigma_2-j\infty}^{\sigma_2+j\infty} F(s)e^{st}\,ds = \int_0^{\infty} f(\tau)\,d\tau\,\frac{1}{2\pi j} \int_{\sigma_2-j\infty}^{\sigma_2+j\infty} e^{s(t-\tau)}\,ds$$

$$= \int_0^{\infty} f(\tau)\delta(t-\tau)\,d\tau$$

$$= f(t) \qquad \text{(A-72)}$$

which proves the inversion formula.

When condition (A-70) is imposed for convergence of the direct transform, then the inversion formula (A-72) will give $f(t)$s which are zero for negative time. To see this we note that if $F(s)$ is a rational fraction,

$$F(s) = \frac{s^n + \cdots + a_1 s + a_0}{s^m + \cdots + b_1 s + b_0} = \frac{(s+s_1)(s+s_3)\cdots}{(s+s_2)(s+s_4)\cdots}$$

$$= \frac{K_2}{s+s_2} + \frac{K_4}{s+s_4} + \cdots \qquad \text{(A-73)}$$

Then
$$f(t) = K_2 e^{-s_2 t} + K_4 e^{-s_4 t} + \cdots \qquad \text{(A-74)}$$

and by (A-70) the contour must lie to the right of all the poles of $F(s)$ to assure convergence—that is, $\sigma_1 > \max\{\text{Re}[s_2], \text{Re}[s_4], \ldots\}$. In this case closure of the contour to the right for $t < 0$ as in Fig. A-11 will enclose no poles and $f(t)$ will be zero for negative time.[1] A more subtle argument can be made leading to the same result when $F(s)$ is not a rational fraction.

While we are on the subject of convergence criteria, we might consider the following question: How is it that we exclude by (A-70) values of s with real parts less than some number σ_1 but at the same time talk about an s plane where all values of s are considered? The answer lies in a technique called *analytic continuation*.[2] We define by (A-68) a function $F(s)$ for values of $\sigma > \sigma_1$. If there is an arc of this boundary where $F(s)$ is analytic, we make a Taylor series expansion about a point on the arc which is valid in a region which was excluded before. In this way we can usually continue the definition of the function over all parts of the s plane in which $F(s)$ is analytic.

A-7. Complex Convolution Formula

Let us now use contour integration to derive the result given in Sec. 8-5 for the transform of a product. We shall first derive the general

---

[1] Evaluation of (A-72) by residues for a rational fraction for $t > 0$ leads to the material in Sec. 7-4.

[2] See, for example, Churchill, *op. cit.*, chap. 11.

expression, paying particular attention to convergence of the integrals involved. We wish to find

$$\mathcal{L}[f_1(t)f_2(t)] = \int_0^\infty f_1(t)f_2(t)e^{-st}\, dt \tag{A-75}$$

The transforms of f_1 and f_2 are

$$\begin{aligned} \mathcal{L}[f_1(t)] &= \int_0^\infty f_1(t)e^{-st}\, dt \qquad \sigma > \sigma_1 \\ \mathcal{L}[f_2(t)] &= \int_0^\infty f_2(t)e^{-st}\, dt \qquad \sigma > \sigma_2 \end{aligned} \tag{A-76}$$

where we have noted that σ, the real part of s, must exceed certain values for convergence. The restriction on σ in (A-75) is

$$\sigma > \sigma_1 + \sigma_2 \tag{A-77}$$

since we require existence of

$$\int_0^\infty |f_1(t)f_2(t)|e^{-\sigma t}\, dt$$

and we know from (A-76) that $f_1(t)$ cannot grow as fast as $\exp(\sigma_1 t)$, nor can $f_2(t)$ grow as fast as $\exp(\sigma_2 t)$. The product, then, cannot grow as fast as $\exp(\sigma_1 t)\exp(\sigma_2 t) = \exp(\sigma_1 + \sigma_2)t$. Now let us write

$$\begin{aligned} \mathcal{L}[f_1(t)f_2(t)] &= \int_0^\infty f_1(t)f_2(t)e^{-st}\, dt \\ &= \int_0^\infty dt\, f_2(t)e^{-st}\frac{1}{2\pi j}\int_{c-j\infty}^{c+j\infty} dw\, F_1(w)e^{wt} \\ &= \frac{1}{2\pi j}\int_{c-j\infty}^{c+j\infty} dw\, F_1(w)\int_0^\infty dt\, f_2(t)e^{-(s-w)t} \\ &\qquad\qquad\qquad\qquad\qquad \mathrm{Re}\,[w] = c > \sigma_1 \end{aligned} \tag{A-78}$$

But

$$\int_0^\infty f_2(t)e^{-(s-w)t}\, dt = F_2(s-w) \qquad \mathrm{Re}\,[s-w] > \sigma_2 \tag{A-79}$$

so that (A-78) becomes

$$\mathcal{L}[f_1(t)f_2(t)] = \frac{1}{2\pi j}\int_{c-j\infty}^{c+j\infty} F_1(w)F_2(s-w)\, dw \tag{A-80}$$

From (A-78) we have

$$\mathrm{Re}\,[w] = c > \sigma_1 \tag{A-81}$$

and from (A-79)

$$\begin{aligned} \mathrm{Re}\,[s-w] &> \sigma_2 \\ \mathrm{Re}\,[s] - \mathrm{Re}\,[w] &> \sigma_2 \\ \sigma - c &> \sigma_2 \\ c &< \sigma - \sigma_2 \end{aligned} \tag{A-82}$$

The final condition, combining (A-81) and (A-82), is

$$\sigma_1 < c < \sigma - \sigma_2 \tag{A-83}$$

which defines a strip in the w plane for the integration of (A-80).

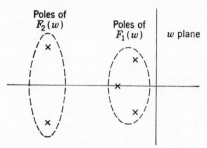

FIG. A-12. Poles of $F_1(w)$ and $F_2(w)$.

As an example, let us suppose that we have $F_1(w)$ and $F_2(w)$ with poles as shown in Fig. A-12. In Fig. A-13 the poles of $F_2(w)$ have been shifted by a change of sign and an arbitrary amount s to correspond to those of $F_2(s - w)$. Also shown are the numbers σ_1, σ_2, etc. We see from (A-83) that the path of integration which lies in the shaded area of Fig. A-14 is

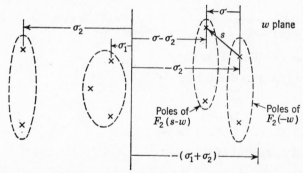

FIG. A-13. Identification of terms for evaluation of Eq. (A-80).

always to the right of the $F_1(w)$ poles and to the left of those of $F_2(s - w)$. If we close the contour to the left, the value of the integral over the semi-circle will vanish if F_1 and F_2 are proper rational fractions, and we have

$$\mathcal{L}[f_1(t)f_2(t)] = \Sigma[\text{residues of } F_1(w)F_2(s - w) \text{ at poles of } F_1(w)] \tag{A-84}$$

Now if $F_1(w)$ has only first-order poles, the residue at $w = s_n$ is by (A-34)

$$k_n = [(w - s_n)F_1(w)F_2(s - w)]_{w=s_n} = F_2(s - s_n)[(w - s_n)F_1(w)]_{w=s_n} \tag{A-85}$$

but the term in brackets on the right is just the residue of $F_1(s)$ alone at $s = s_n$. If we call this residue b_n, then the result is

$$\mathcal{L}[f_1(t)f_2(t)] = \sum_n b_n F_2(s - s_n) \tag{A-86}$$

which confirms (8-36).

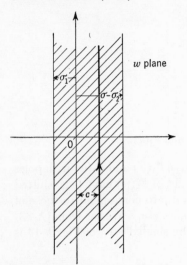

FIG. A-14. Region for path of integration of Eq. (A-80).

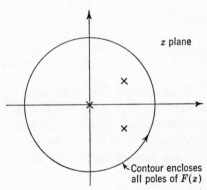

FIG. A-15. Contour for inversion of Z transform.

A-8. Inversion of the Z Transform

The Z transform as defined by (16-13) is

$$\mathcal{Z}[f(n)] \triangleq \sum_{n=0}^{\infty} f(n)z^{-n} \triangleq F(z) \tag{A-87}$$

We write out the terms of the sum:

$$F(z) = f(0) + \frac{f(1)}{z} + \cdots + \frac{f(k-1)}{z^{k-1}} + \frac{f(k)}{z^k} + \frac{f(k+1)}{z^{k+1}} + \cdots \tag{A-88}$$

Now this expansion will converge[1] for all values of z such that $|z|$ exceeds the distance from the origin to the furthest pole of $F(z)$. This is analogous to saying the Laplace integral converges for all values of s to the right of the pole with greatest real part. Later we can extend the region of validity to the whole z plane by analytic continuation as discussed in Sec. A-6.

[1] This is a special case of *Laurent's series*. See, for example, Churchill, *op. cit.*, p. 102.

Let us now multiply (A-88) by z^{k-1} and integrate around a contour which encloses all poles of $F(z)$ as shown in Fig. A-15. The integral is

$$\oint F(z)z^{k-1}\,dz = \oint dz\left[f(0)z^{k-1} + f(1)z^{k-2} + \cdots\right.$$
$$\left. + f(k-1) + \frac{f(k)}{z} + \frac{f(k+1)}{z^2} + \cdots\right] \quad \text{(A-89)}$$

Now the residue of the integrand on the right is the coefficient of the z^{-1} term, which is $f(k)$. Then we have the result

$$\frac{1}{2\pi j}\oint F(z)z^{k-1}\,dz = f(k) \quad\quad\quad \text{(A-90)}$$

which corresponds to (16-15).

PROBLEMS

A-1. Show that if

$$z(t) = e^{jt}$$

the path traced in the z plane is the unit circle. Plot this, and show points for $t = \pi/2,\ \pi$.

A-2. Plot the path on the z plane for

    a. $z = t$
    b. $z = jt$
    c. $z = t(1 + j)$

A-3. Show that $|z|$ is not analytic.

A-4. Test for analyticity at $z = 0$

    a. $f(z) = \exp\left(-\dfrac{1}{z}\right)$
    b. $f(z) = e^{-z}$
    c. $f(z) = \dfrac{1}{z+1}$

A-5. Show that

$$\int_C \frac{z-3}{z}\,dz = 4 + 3j\pi$$

PROB. A-5

A-6. Show by direct integration that

$$\oint \frac{z-3}{z}\,dz = -6\pi j \qquad C\colon |z| = 2$$

A-7. Find the residues of the following functions at their poles.

a. $\dfrac{z+1}{z^2-2z}$ *Ans.* $-\frac{1}{2}, \frac{3}{2}$

b. $\dfrac{z^3}{(z+3)^2}$

c. $\dfrac{z}{(z+1)(z+3)}$ *Ans.* $-\frac{1}{2}, \frac{3}{2}$

A-8. Solve (A-6) by the theory of residues.

A-9. Find

$$\oint_C \frac{5z-2}{z(z-1)}\,dz \qquad C: |z| = 2 \qquad\qquad Ans.\quad 10\pi j$$

A-10. Find for C a unit circle

a. $\oint_C \dfrac{e^{-z}}{z^2}\,dz$ *Ans.* $-2\pi j$

b. $\oint_C \dfrac{e^z}{z^3}\,dz$ *Ans.* πj

A-11. Find by complex convolution the following.   Make a plot of poles, and locate the path of integration.

   a. $\mathcal{L}[e^{-at} \cdot e^{-bt}]$
   b. $\mathcal{L}[e^{-at} \cdot t]$
   c. $\mathcal{L}[e^{-\alpha t} \cos \beta t]$

A-12. Give a proof for (A-36).

TABULATION OF £ TRANSFORMS

B-1. £ Transforms of Operations and Functions

Table B-1 contains the operation pairs derived in Chaps. 2 and 8.

TABLE B-1†

$f(t)$	$F(s)$	Section
$\dfrac{df}{dt}$	$sF(s) - f(0)$	2-2
$\dfrac{d^2f}{dt^2}$	$s^2F(s) - sf(0) - f'(0)$	2-2
$\dfrac{d^nf}{dt^n}$	$s^nF(s) - s^{n-1}f(0) - \cdots - f^{(n-1)}(0)$	2-2
$\displaystyle\int_0^t f(x)\,dx$	$\dfrac{F(s)}{s}$	2-3
$f(t - a)u(t - a)$	$e^{-as}F(s), \qquad a \geq 0$	8-2
$f(t + a)u(t)$	$e^{as}\mathcal{L}[f(t)u(t-a)], \qquad a \geq 0$	8-2
$g(t)$ (periodic, period T)	$\dfrac{G_1(s)}{1 - e^{-Ts}}$ $G_1(s) = \mathcal{L}[g(t)u(t - T)]$	8-2
$f(t)e^{-at}$	$F(s + a)$	8-3
$f(at)$	$\dfrac{1}{a}F\left(\dfrac{s}{a}\right) \qquad a \geq 0$	8-4
$f_1(t)f_2(t)$	$\dfrac{1}{2\pi j}\displaystyle\int_{c-j\infty}^{c+j\infty} F_1(w)F_2(s - w)\,dw$	8-5, A7
$\displaystyle\int_0^t f_1(\tau)f_2(t - \tau)\,d\tau$	$F_1(s)F_2(s)$	8-6
$\dfrac{\partial f(t,a)}{\partial a}$	$\dfrac{\partial F(s,a)}{\partial a}$	8-7
$\displaystyle\int_{a_1}^{a_2} f(t,a)\,da$	$\displaystyle\int_{a_1}^{a_2} F(s,a)\,da$	8-7
$t^nf(t)$	$(-)^n\dfrac{d^n}{ds^n}F(s)$	8-8
$\dfrac{f(t)}{t}$	$\displaystyle\int_s^\infty F(s)\,ds$	8-8
$\text{Re}\,[f(t)]$	$\text{Re}\,[F(s)]$	8-9
$\text{Im}\,[f(t)]$	$\text{Im}\,[F(s)]$	8-9
$f(0) = \lim\limits_{s \to \infty} sF(s)$		8-10
$f(\infty) = \lim\limits_{s \to 0} sF(s)$ (poles of $sF(s)$ in left half of s plane)		8-10

† For other properties, see A. Erdélyi, "Tables of Integral Transforms," vol. 1, McGraw-Hill Book Company, Inc., New York, 1954.

TABLE B-2. DIRECT $\mathcal{L}$ TRANSFORMS

$f(t)$	$F(s)$
$\delta_+(t)$	1
$\delta'_+(t)$	s
$u(t)$	$\dfrac{1}{s}$
t	$\dfrac{1}{s^2}$
t^2	$\dfrac{2}{s^3}$
t^n	$\dfrac{n!}{s^{n+1}}$
e^{-at}	$\dfrac{1}{s+a}$
te^{-at}	$\dfrac{1}{(s+a)^2}$
$t^n e^{-at}$	$\dfrac{n!}{(s+a)^{n+1}}$
$\sin \beta t$	$\dfrac{\beta}{s^2 + \beta^2}$
$\cos \beta t$	$\dfrac{s}{s^2 + \beta^2}$
$\sinh \beta t$	$\dfrac{\beta}{s^2 - \beta^2}$
$\cosh \beta t$	$\dfrac{s}{s^2 - \beta^2}$
$e^{-\alpha t} \sin \beta t$	$\dfrac{\beta}{(s+\alpha)^2 + \beta^2}$
$e^{-\alpha t} \cos \beta t$	$\dfrac{s+\alpha}{(s+\alpha)^2 + \beta^2}$
$t \sin \beta t$	$\dfrac{2\beta s}{(s^2 + \beta^2)^2}$
$t \cos \beta t$	$\dfrac{s^2 - \beta^2}{(s^2 + \beta^2)^2}$
$te^{-\alpha t} \sin \beta t$	$\dfrac{2\beta(s+\alpha)}{[(s+\alpha)^2 + \beta^2]^2}$
$te^{-\alpha t} \cos \beta t$	$\dfrac{(s+\alpha)^2 - \beta^2}{[(s+\alpha)^2 + \beta^2]^2}$

TABLE B-3. $\mathcal{L}$-TRANSFORM PAIRS DERIVED IN THE TEXT

$f(t)$	$F(s)$	Section
t^ν	$\dfrac{\Gamma(\nu + 1)}{s^{\nu+1}}$	2-4
erf $(t)^{1/2}$	$\dfrac{1}{s\sqrt{s+1}}$	2-4
$\dfrac{4}{\pi}\displaystyle\sum_{n=0}^{\infty}\dfrac{\sin(2n+1)t}{2n+1}$	$\dfrac{1}{s}\tanh\dfrac{\pi s}{2}$	8-2
$u(t-1)$	$\dfrac{e^{-s}}{s}$	8-4
$\dfrac{\sin t}{t}$	$\tan^{-1}\dfrac{1}{s}$	8-11
$\dfrac{\sin^2 \beta t}{t}$	$\dfrac{1}{4}\ln\left[\dfrac{s^2+4\beta^2}{s^2}\right]$	8-7
$L_n(t)$	$\dfrac{(s-1)^n}{s^{n+1}}$	8-10
$J_0(t)$	$\dfrac{1}{\sqrt{s^2+1}}$	8-10
Si(at)	$\dfrac{1}{s}\tan^{-1}\dfrac{a}{s}$	8-11

B-2. Inverse $\mathcal{L}$ Transforms

Table B-4 gives inverse transforms of rational fractions with one, two, and three poles. The inverse transforms are given in terms of s-plane geometry as discussed in Secs. 2-5 and 7-6.

In the table, lengths are measured positive to the right as shown by the arrows. When a length is designated by a line with arrows on both ends, it is always taken as positive.

TABLE B-4

One Pole

$F(s)$	s plane	$f(t)$
$\dfrac{1}{s+a}$		e^{-at}

Two Poles

$\dfrac{1}{(s+a_1)(s+a_2)}$		$\dfrac{1}{A}(e^{-a_1 t} - e^{-a_2 t})$
$\dfrac{(s+b)}{(s+a_1)(s+a_2)}$		$\dfrac{1}{A}(B_1 e^{-a_1 t} - B_2 e^{-a_2 t})$
$\dfrac{1}{(s+a)^2}$		te^{-at}
$\dfrac{(s+b)}{(s+a)^2}$		$(Bt+1)e^{-at}$
$\dfrac{1}{(s+a)^2 + \beta^2}$		$\dfrac{1}{\beta} e^{-at} \sin \beta t$
$\dfrac{s+b}{(s+a)^2 + \beta^2}$		$\dfrac{B}{\beta} e^{-at} \sin (\beta t + \theta)$

TABLE B-4 (*Continued*)

	Three Poles	
$F(s)$	s plane	$f(t)$
$\dfrac{1}{(s+a_1)(s+a_2)(s+a_3)}$		$\dfrac{1}{A_{21}A_{31}}\,e^{-a_1 t} - \dfrac{1}{A_{21}A_{32}}\,e^{-a_2 t} + \dfrac{1}{A_{31}A_{32}}\,e^{-a_3 t}$
$\dfrac{s+b}{(s+a_1)(s+a_2)(s+a_3)}$		$\dfrac{B_1}{A_{21}A_{31}}\,e^{-a_1 t} - \dfrac{B_2}{A_{21}A_{32}}\,e^{-a_2 t} + \dfrac{B_3}{A_{31}A_{32}}\,e^{-a_3 t}$
$\dfrac{s^2}{(s+a_1)(s+a_2)(s+a_3)}$		$\dfrac{a_1{}^2}{A_{21}A_{31}}\,e^{-a_1 t} - \dfrac{a_2{}^2}{A_{21}A_{32}}\,e^{-a_2 t} + \dfrac{a_3{}^2}{A_{31}A_{32}}\,e^{-a_3 t}$
$\dfrac{1}{(s+a_1)^2(s+a_2)}$		$\dfrac{1}{A^2}\,[e^{-a_2 t} + (At-1)e^{-a_1 t}]$
$\dfrac{s+b}{(s+a_1)^2(s+a_2)}$		$\dfrac{1}{A^2}\,[B_2 e^{-a_2 t} + (AB_1 t - B_2)e^{-a_1 t}]$
$\dfrac{s^2}{(s+a_1)^2(s+a_2)}$		$\dfrac{1}{A^2}\,[a_2{}^2 e^{-a_2 t} + (a_1{}^2 At - a_1 A - a_1 a_2)e^{-a_1 t}$
$\dfrac{1}{(s+a_2)[(s+\alpha)^2 + \beta^2]}$		$\dfrac{1}{A^2}\,e^{-a_2 t} + \dfrac{1}{A\beta}\,e^{-\alpha t}\sin(\beta t - \theta)$

TABLE B-4 (*Continued*)

Three Poles (*Continued*)

$F(s)$	s plane	$f(t)$
$\dfrac{s + b}{(s + a_1)[(s + \alpha)^2 + \beta^2]}$		$\dfrac{B_1}{A^2} e^{-a_1 t} + \dfrac{B}{\beta A} e^{-\alpha t} \sin(\beta t - \theta)$
$\dfrac{s^2}{(s + a_1)[(s + \alpha)^2 + \beta^2]}$		$\dfrac{a_1{}^2}{A^2} e^{-a_1 t} + \dfrac{B^2}{\beta A} e^{-\alpha t} \sin(\beta t - \theta_1 - 2\theta_2)$

INDEX